Killester

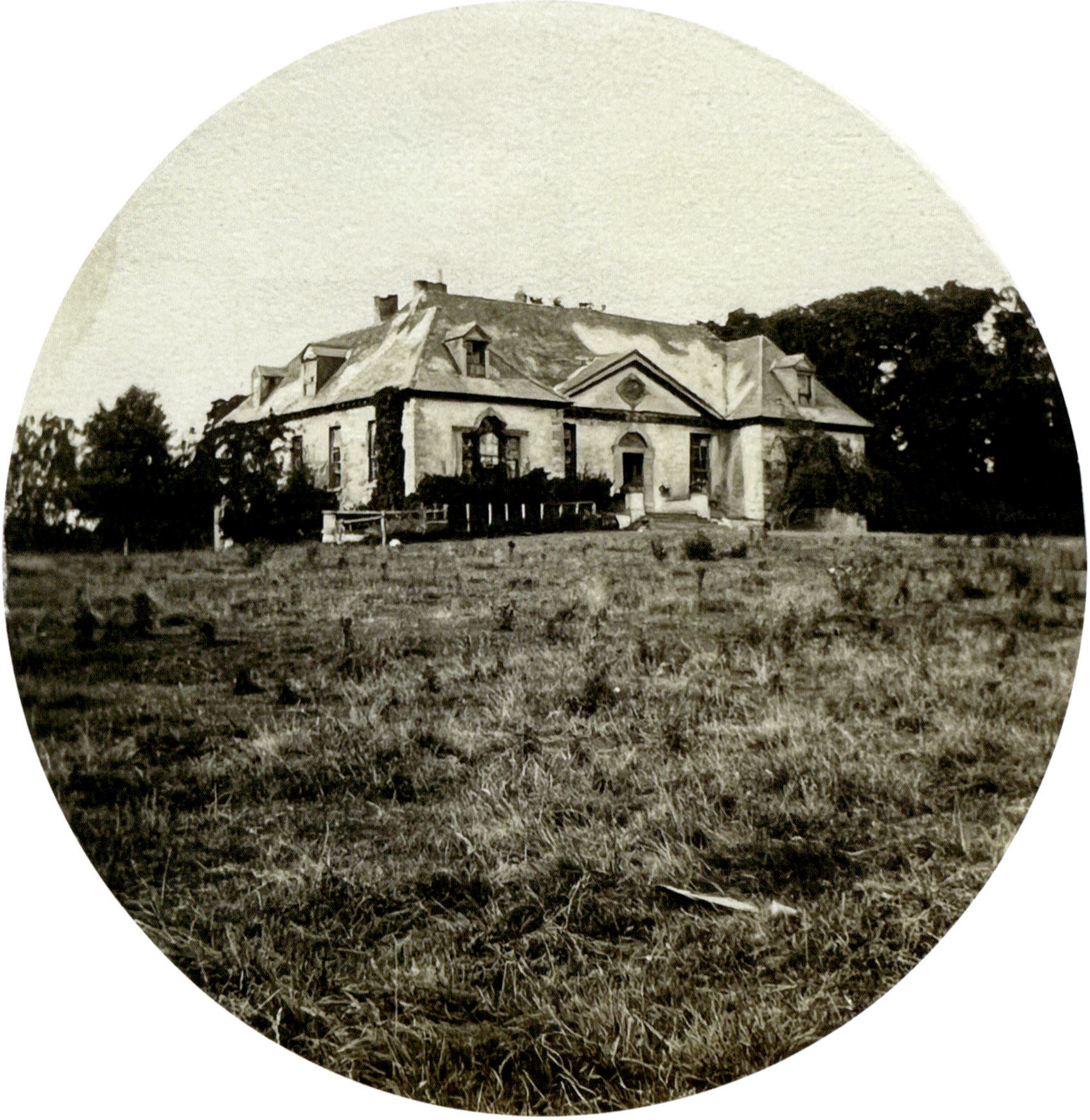

DUBLIN – SHAPING THE SUBURBS

Charles Duggan, SERIES EDITOR

Killester

From medieval manor to garden suburb

Joseph Brady and Ruth McManus

with archival research by
Abigail O'Reilly

FOUR COURTS PRESS
in association with
DUBLIN CITY COUNCIL

Typeset in 10.5pt on 13.5pt AdobeGaramondPro by
Carrigboy Typesetting Services for
FOUR COURTS PRESS LTD
7 Malpas Street, Dublin 8, Ireland
www.fourcourtspress.ie
and in North America for
FOUR COURTS PRESS
c/o IPG, 814 N Franklin St., Chicago, IL 60610.

© the various authors and Four Courts Press 2025

A catalogue record for this title is available from the British Library.

ISBN 978-1-80151-193-3

Page ii shows an image of Killester House, reproduced with the kind permission of the Gaisford-St Lawrence family collection.

All rights reserved. Without limiting the rights under copyright reserved alone, no part of this publication may be reproduced, stored in or introduced into a retrieval system, or transmitted, in any form or by any means (electronic, mechanical, photocopying, recording or otherwise), without the prior written permission of both the copyright owner and the publisher of this book.

SPECIAL ACKNOWLEDGMENT

This book is published in association with Dublin City Council. It is an action of the Dublin City Strategic Heritage Plan 2024–9, and part funded by the Heritage Council Heritage Plan Grant Scheme.

Printed in Poland
by L&C Printing Group, Krakow.

Contents

LIST OF ABBREVIATIONS — vi

ACKNOWLEDGMENTS — vii

INTRODUCTION — ix
Charles Duggan

CHAPTER 1 The evolution of Killester from earliest times — 1
Ruth McManus

CHAPTER 2 The Killester Garden Village — 46
Joseph Brady

CHAPTER 3 The suburbanization of Killester in the twentieth century — 99
Ruth McManus

NOTES — 126

BIBLIOGRAPHY — 127

LIST OF ILLUSTRATIONS — 131

INDEX — 135

Abbreviations

DIB	*Dictionary of Irish biography*
HMSO	His/Her Majesty's Stationery Office
HM Treasury	His/Her Majesty's Treasury Department
ISSLT	Irish Sailors' and Soldiers' Land Trust
LGB	Local Government Board for Ireland
MP	Member of parliament
NAI	National Archives of Ireland
NLI	National Library of Ireland
OPW	Office of Public Works
PLV	Poor Law Valuation
RCSI	Royal College of Surgeons in Ireland
RDS	Royal Dublin Society
TCD	Trinity College Dublin
TD (Teachta Dála)	Member of Dáil Éireann, the lower house of the Irish parliament
UCD	University College Dublin

UNITS

Area (UK statute)

Acres (*a.*)	0.4 hectares
Rood (*r.*)	1,012 sq. metres
Perch (*p.*)	25.3 sq. metres
Square foot	0.092 sq. metres

There are four roods to an acre, and a rood contains 40 perches.

Distance (UK statute)

Mile	1.6 kilometres
Yard	0.91 metres
Foot	30.5 cms
Inch	2.54 cms

Irish Measure

Irish mile	1.27 statute miles
Irish (plantation) acre	1.62 statute acres

Currency

A pound contained four crowns, twenty shillings (*s.*) or 240 pennies (*d.*). A guinea was twenty-one shillings. Reference prior to Irish independence is to the 'pound sterling', the currency in force in the United Kingdom. Following independence, an Irish equivalent of the UK currency was created with a value at par until March 1979 when an exchange rate was introduced.

Acknowledgments

This book would not have been possible without the support and assistance of Dublin City Council. We wish to thank Charles Duggan (Heritage Officer of Dublin City Council) for commissioning this important series and Abigail O'Reilly, Graduate Heritage Officer, whose research ensured that we have many superb images, which we otherwise would not have been able to use. A desk-based archaeological survey undertaken by Antoine Giacometti and Sam Hughes (Archaeology Plan Ltd) for Dublin City Council provided very useful data which we have incorporated into the various chapters. We would like to thank Colm Lennon for his helpful advice on sources, particularly in connection with the St Lawrences and Vernons in Clontarf.

We are indebted to the North Central Area Committee for their support and also to Mick Carroll and Sheila Duffy from the North Central Area Office who part funded this publication.

We wish to acknowledge the support of Dublin City Council's Chief Executive Richard Shakespeare, Deirdre Scully, City Planner, Malachy Bradley, Deputy City Planner, Máire Igoe, Acting Executive Manager, and Rhona Naughton, Senior Planner for Conservation and Heritage. We also acknowledge the support of Dr Ruth Johnson and the Dublin City Council Archaeology team.

Sincere thanks to Lorraine McLoughlin and the staff of Dublin City Archives, James Harte and the staff of the National Library of Ireland, Fingal County Archivist Karen de Lacey and Catherine Keane, whose help with the Fingal Local Studies and Archives was essential; Vincent Buttner and the staff of the National Archives of Ireland; the staff of UCD Digital Library; Heather Holmes from Freeman's Auctioneers; Harriet Hunter-Smart from Tennants Auctioneers; James Grange Osborne from Independent Archives; Mindy Lynch from the Historic Environment Scotland Archives; Jennifer Smith from the National Trust; Paul Johnson from the UK National Archives; Daniel Eglinton-Carey for his assistance in developing our understanding of the Howth estate and to Douglas Appleyard of the Raheny Heritage Society who provided background information on the Killester Demesne from the society's records. We are also very grateful to the Glucksman Map Library in Trinity College and to Paul Ferguson, the map librarian, for assistance with many images.

We are also grateful to the individuals and organisations who licenced or gave permission to reproduce the images that appear in this text. This is noted with each individual image and in the list of images at the back.

This book is an action of the Dublin City Strategic Heritage Plan 2024–9 and received County Heritage Plan Grant Funding from the Heritage Council.

Introduction

CHARLES DUGGAN, DUBLIN CITY HERITAGE OFFICER

Killester is a small suburban area situated in north Dublin on the Howth Road between Clontarf and Raheny. It contains within its boundaries a rich and varied history that stretches back over centuries. This book, *Killester: from medieval manor to garden suburb*, unpacks that history, offering an in-depth exploration of Killester from its medieval origins to its transformation into a garden suburb for ex-servicemen returning from the First World War. It provides insights into the key historical figures, families, and developments that shaped the area into what it is today. The book was commissioned to mark the centenary of this garden suburb and to contribute to the on-going celebrations of and research on Killester championed by the Killester Garden Village Committee, whose book, *Killester Garden Village: the lives of Great War veterans and their families* (Dublin), was published in 2024.

At the heart of Killester's early history is its connection to the Anglo-Norman settlement in Dublin. Ruth McManus writes about how the area first came under the influence of William Brun, a follower of King Henry II, who was granted the land of Killester in the late twelfth century. This early period laid the foundations of Killester's development as a manorial estate. Over time, ownership shifted to the White family and eventually to the St Lawrence family, the lords of Howth, solidifying Killester's place in the medieval landscape of Dublin.

Killester's role as a manorial centre is evident in its now built over agricultural lands, medieval church and graveyard, and, possibly, a long-vanished fortified residence about which the records are virtually silent. By the late seventeenth or early eighteenth century Killester House was built. Attributed to Thomas Burgh, surveyor general, this was a substantial dormer single-storey over basement house in which a succession of influential residents lived including Sir William Gleadowe Newcomen, and his successors, who undertook great improvements to the estate.

The twentieth century marked a radical transformation for Killester. Following the First World War, the UK government embarked on a housing initiative to provide homes for war veterans. Killester was selected as the site for a garden suburb, part of a wider 'homes fit for heroes' initiative across Britain and Ireland. The garden suburb model, influenced by the ideas of Ebenezer Howard, aimed to combine the benefits of the city with the peace and greenery of the countryside, providing a well-planned, healthy living environment for working-class families.

Joseph Brady writes in detail about the complex story of the development of Killester as a garden suburb for ex-servicemen and their families, its subsequent management by the Irish Sailors' and Soldiers' Land Trust and the chequered relationship the Trust had with its Killester tenants. He explains how the scheme helped to address the urgent housing needs of post-war Ireland but also how the scheme stands alone in the suburbanization of Dublin in terms of its protracted and complex development history. The Killester scheme, and its near contemporary in Marino, which are celebrating their centenaries, are important early Irish examples of this new vision for urban planning.

As the decades passed, Killester continued to grow and evolve. The 1930s saw further housing developments and as the community expanded new shops, services, and schools catering to the increasing population emerged. By the mid-twentieth century, Killester had become a vibrant suburb. Despite these changes, important vestiges of old Killester can still be seen there today including the medieval church and graveyard, stretches of rubble stone boundary walls and eighteenth-century redbrick garden walls and possibly the archaeological remains of Killester House itself. This book complements other Dublin City Council-sponsored initiatives, such as the Irish Historic Towns Atlas Dublin Suburbs Series, which has already published volumes on Clontarf and Rathmines, with three more in progress for Drumcondra, Kilmainham-Inchicore, and Ringsend.

ABOUT THE SERIES

This richly illustrated book is the first in a limited series on Dublin's twentieth-century suburbs commissioned by Dublin City Council under the auspices of the Dublin City Strategic Heritage Plan 2024–9. Other volumes are in preparation and will appear in the near future. This series aims to explore Dublin's suburbs as they reach their centenary years, seeing them as layered and complex places, and more than simply the backdrop to the city's twentieth-century expansion, and celebrating them as dynamic places where the past and present converge, revealing how community resilience, urban planning ambitions, and socio-political forces have shaped them over time.

CHAPTER ONE

The evolution of Killester from earliest times

RUTH McMANUS

At first glance, the contemporary visitor to Killester might consider that there is little of historical interest to note in this suburban locality. However, despite a limited number of above ground archaeological traces, the area can boast of a rich heritage. Its street names, layouts and occasional glimpses of historic building fabric are vestiges of a long history that can be traced back to at least the twelfth century. This chapter examines the evidence for that history from earliest times until the destruction of Killester House in the early twentieth century. It is not intended as a definitive account, but as a starting point from which others may build in order to develop a deeper understanding of this intriguing locale.

PLACENAME ORIGINS OF KILLESTER / CILL EASRA

The modern-day placename of Killester or Cill Easra refers to a suburb, which is also a civil parish in the barony of Coolock. Another Killester townland is located near Ratoath in Co. Meath. National placenames database *logainm.ie* records various usages from 1178 to 1489, with variations including Cellesra, Killesra, Kilastra, Cellestra, Killestre, Killestry. The first spelling as Killester is recorded in 1504 and thereafter this is the dominant usage, although Kilaster is noted at least once. Garrett (2006: 5) notes that Archbishop Laurence O'Toole refers to Quilesra in the twelfth century, while King John called it Tudressa. He suggests that it has also been called Kylestre, Kyllester and Xyllester.

While the first part of the Irish name is readily understood, with 'cill' being a common placename element referring to a church, the second part is less straightforward. The modern usage is 'Cill Easra', the church of Easra, but this appears to be relatively recent. In 1837 the Ordnance Survey parish namebooks explained the Irish language version as follows: 'Cill Laistre, Laisre, St Lassera's church'. A 2005 report on the graveyard claims that the ancient church was originally dedicated to Saint Lasera but subsequently dedicated to Saint Brigid sometime between the fifth and ninth centuries, as 'according to legend St Brigid is said to have performed several miracles in Killester during her visit to the convent of Lasera' (Natura Environmental Consultants, 2005: 7). Despite this assertion, the evidence is far from convincing. Saint Lasair is a somewhat obscure female saint, reputedly one of Saint Rónán's six

I

daughters, and associated with Killesher (Cill Lasair), Co. Fermanagh and Kilronan in Co. Roscommon. A brief life, *Beatha Lasrach*, compiled in the seventeenth century, was translated by Lucius Gwynn in 1911. A more recent study by Seosamh Ó Dufaigh (2004) places the composition of the story of St Lasair's life in the north Connacht / south Ulster region in the mid- to late medieval period. There is no connection to Dublin or the east coast of Ireland more generally. Revd James Kenny, who wrote a history of the parish of Coolock in 1934, also noted the lack of a Killester connection to St Lasair. Placenames associated with St Lasair are generally anglicized using the suffix -esher, as in Killesher in Co. Fermanagh, rather than -ester. Kenny proposed an alternative origin for the name, based on the old Irish term 'leaster', which refers to a vessel or container for liquid (Kenny, 1934: 8–9). If correct, the name Killester could originate from the term 'church of the vessel / chalice / container'.

The modern Irish translation of Killester as Cill Easra, meaning the church of Easra, has been used to inform the naming of the redeveloped former Killester Cinema site on Collins Avenue as Easra Court. Although a recent infill development off Killester Avenue is named St Esra's Close, *logainm.ie* observes that there is no evidence for the existence of a saint with that name.[1] The School's Folklore collection, recorded in the 1930s, includes a rather unlikely account of how Killester got its name, linking it to Cromwellian soldiers attempting to kill a nun named Ester while fleeing the local convent.[2] This is interesting as there is a reference here to a convent, which may link to the local name 'Nun's Walk', despite the fact that there is no documentary evidence for such an institution in the locality during the Cromwellian era (1649 to 1653). A 'convent in ruins' is depicted on the first Ordnance Survey map of the locality from the 1830s, located on the Howth Road close to the present-day Saint Brigid's church, but no further information is currently available. To complicate matters further, the current Roman Catholic parish church, consecrated in 1926, is dedicated to Saint Brigid, and this name was selected based on the belief that there had been a small chapel in the locality with that dedication as early as the Viking era. The only statement that can be made with certainty is that Killester has been the location of a Christian place of worship for more than a millennium.

EARLIEST HISTORICAL REFERENCES TO KILLESTER

Some of the earliest documented evidence for Killester (referring variously to Killastre and to Quillestre) is in the form of charters dating to the twelfth century. These charters were legal documents created between 1178 and 1180 to record the granting of lands in the locality. At the time charters were a relatively new feature in Ireland; it is thought that they came into use when the Cistercian order first arrived in the country in the 1140s. The charters relating to Killester survived in the possession of the lords of Howth and are now held in the National Library of Ireland.

The evolution of Killester from earliest times

1 Feoffment from Jenico (Preston) Viscount Gormanston and others, to Nicholas (St Lawrence) Lord Howth, of the lands of Killester, Baldongan, and other townlands in Co. Dublin, 1626 May 1. [D.9994 Howth Castle Papers. NLI]

> Be it known to all both now and in the future that I Laurence, by the grace of God archbishop of Dublin, have conceded, and by this charter confirmed, to William Brun and his heirs the land of Quillesra which the canons of the church of the Holy Trinity granted to them and by counsel and agreement of the whole chapter confirmed with their charter, to have and to hold of the same church in fee and hereditary succession, with all its appurtenances, freely and peacefully and honourably. In land, in sea, in wood, in plain, in meadows, in pastures, in roads, in paths, and in all liberties pertaining to the same land, for the annual payment on the feast-day of St Michael on the altar of Holy Trinity of half an ounce of gold and boots to the prior of the same church and a full tenth of the land. Witnessed by: Gervase the prior, William Fitzaldelin, king's steward, Adam, abbot of St Mary's, Christinus the sacristan, Radolphus, priest of St John's, Aelamus brother of Haim, Gilbert Gudo, Thomas de Schiterbi, Walter the goldsmith, Richard of York and Gilbert the clerk who wrote this charter.
> Transcription of the Killester Charter (Garrett, 2006: 4)

In the twelfth century, much of the area of county Dublin was in the ownership of the archbishop of Dublin and of various monastic houses. The priory of the Holy Trinity (Christ Church cathedral), which had been founded about 1038 by King Sitric of Dublin, was the wealthiest religious house in Ireland. It held over 10,000 acres of property in Co. Dublin alone, including Glasnevin, Raheny, Killester and Clonturk (modern-day Drumcondra) (Gillespie, 1996). Nearby Kilbarrack and Donnycarney were owned by St Mary's abbey, which had been founded about 948 (Walsh, 1888).

We know from a series of charters that the prior of Christ Church had already been the lord of Killester before the advent of the Anglo-Normans. Now possession of the land was granted to one of the new arrivals, William Brun, a follower of King Henry II. The fee farm grant arrangement was similar to a freehold but required payment of an annual rent. In this case the payment consisted of a half an ounce of gold and a pair of boots for the prior to be presented on the altar of the church of Holy Trinity at Michaelmas, 29 September (Mills, 1891: 26 and 164).[3] By 1343, five shillings was received for the rental for Killester for a whole year, according to an account roll of the priory. Garrett (2006: 6) claims that Gervase, the first prior of Christ Church, was intimidated and forced to demise the lands of Killester to the adventurer William Le Brun, while Liam Howlett's (1979) article on the early charters notes that William Brun, together with his wife Susanne, son Audoen and daughters Alice and Susanne probably lived in Killester.

Although we do not know for certain what the residence of the Brun family was like, the balance of probability points to there having been a motte and bailey castle at Killester at this time. This is a typical Anglo-Norman fortification consisting of either a wooden or stone keep located on a raised area of ground (motte) with a walled courtyard (bailey), surrounded by an earthen rampart and ditch. Father James Kenny's *History of Coolock parish* (1934: 10) claims that a 'remarkable earthwork' that once surrounded Killester House had survived into the twentieth century. He described this as a five-foot high embankment with an exterior ditch which could still be seen in locations where construction had not yet taken place. Kenny attributed the earthwork to the Anglo-Normans because it had angles and corners, unlike the style of earlier circular Gaelic ringforts. Garrett (2006: 11) also claims that 'the outer boundaries of the estate were secured by a medieval fosse or moat, topped by a wickerwork fence'. Murphy and Potterton (2010) have further suggested that the Brun family may have built the motte 'that has since been destroyed and whose exact location is a matter of speculation.' There is a distinct scarcity of motte and baileys in the immediate Dublin hinterland, as Stout (2012) has observed. Indeed, if such a structure did exist at Killester, it is the only known example of a motte on church lands in Dublin (Murphy and Potterton, 2010). Perhaps reflecting the need for such defensive structures, Brun was murdered in 1199 (Howlett, 1979: 71).

As a medieval manorial centre, in addition to the residence of the lord of the manor, Killester would have had various agricultural buildings and other features, which could have included mills, barns, granaries, dovecots and warrens. Typically, the manorial centre was situated close to the parish church, as appears to have been the

The evolution of Killester from earliest times

The Church at KILLESTER. 2½ M. from Dublin.

2 Killester church (in ruins), 1769. Gabriel Beranger. [PD 1958 TX_001(B). NLI]. A second, very similar, illustration by Beranger and dating to 1769 is also available in the National Library of Ireland, PD 2100 TX 1(A).

case at Killester, if the sites of the ruined church and of the mansion house as depicted on the first edition Ordnance Survey map represent continuity of site and function from the medieval period. As noted above, Garrett (2006: 5–6) claims that there was a church or small chapel at Killester dedicated to Saint Brigid dating back to 'the years of Norse settlements' which was originally appendant to the monastery at Swords but later transferred to the priory of the Holy Trinity (Christ Church). By the 1650s, the Civil Survey referred to the walls of a decayed chapel at this location. The National Monuments Service (*archaeology.ie*) considers the rectangular building at the centre of the graveyard on the north side of Killester Avenue to be the surviving remnants of the medieval parish church of Killester. This plain building is typical of the many medieval churches in north Dublin, including those at Kilbarrack graveyard and Grange Abbey,

Donaghmede. According to D'Alton (1838) the church was still standing but unroofed in 1838. Murray's (2009) discussion of Killester teases out the changing position of the chapel in the period from the early sixteenth to the mid-seventeenth century. Whereas Alen's Register described it as a chapel in c.1505 and c.1530, after the alteration of Christ Church into a secular cathedral, Killester was granted to the economy and was thereafter described as a piece of property. Together with the fact that it does not appear in any of the early seventeenth-century visitations, this implies that Killester did not function as a parochial entity after that date. The chapel had decayed and gone out of use, with this 'decayed chapel' being noted in the mid-seventeenth-century Civil Survey. Once the church went out of use it would have been unroofed, although burials continued long after this date, with the graveyard finally being closed in 1876. The ruinous church at Killester attracted artistic interest, with surviving watercolours dating from the 1760s and 1770s. By the late nineteenth century, Walsh (1888: 225) describes it as having been a 'very plain building, of roughly hammered stone'. The ruins were hidden beneath elder trees and a high stone wall shut the ruins from public view. Both east and west gables were still standing, with a window, but no belfry remained. Each gable had a window and the south wall had a low-arched doorway, whereas the north wall was very broken. This description fits with the watercolours attributed to Gabriel Beranger from the previous century. Another very attractive sketch of the ruinous church and graveyard dating to 1826 is held by the National Gallery of Ireland and attributed to William Howis Sr.

KILLESTER AND THE LORDS OF HOWTH

It is unclear exactly when the manor of Killester passed out of the Brun family, but by the fourteenth century the White family was in control of Killester (Howlett, 1979). Garrett (2006) has suggested that the White family, original owners of Corr Castle at Howth, intermarried with the Bruns and had become prominent in the Anglo-Norman government at the time they came into possession of the Killester lands. As early as 1373, one Richard White of Killester was recorded as a member of a commission of the peace, while in the following century a Robert White of Killester fulfilled similar roles in 1423 and 1425 (Frame, 1992). Sources differ as to the exact date, but somewhere between 1459 and 1465, Alice White, daughter of Nicholas White of Killester, married Robert St Lawrence, second baron of Howth (*DIB*; Garrett, 2006: 7).[4] Through this marriage, the lands and manor of Killester came into the possession of the Howth family. The connection between the lords of Howth and Killester would persist for well over 500 years.

The main residence at Killester must have been relatively significant and of sufficient standing to be used as a dwelling by various members of the Howth family over the years (Armstrong, 1915: 143; Ball, 1917; Dawson, 1976: 127). According to

The evolution of Killester from earliest times 7

3 Extract from map of the County Dublin. William Petty, *Hiberniae delineatio*, 1683, showing 'Killaster' placename. [GE CC 1260. BNF]

the Book of Howth, Nicholas St Lawrence – third baron of Howth and son of Robert and Alice White – entertained Sir James Butler at his mother's house in Killester in 1492 (Bowen, 1963: 74). At this time the Geraldines and the Butlers had been famously reconciled to the extent that Sir James Butler and the earl of Kildare had shaken hands through an opening in a closed door of St Patrick's cathedral. However, at the dinner in Killester, Nicholas St Lawrence apparently resented a verbal attack that Butler made on the earl of Kildare and challenged him to a duel. Butler did not accept the challenge and left in a fury (Ball, 1917). By the time of his death in 1526, Nicholas St Lawrence held from the priory [of the Holy Trinity, or Christ Church] the manor of Killester, at the yearly rent of 3s. 4d. (*Repertorium Viride*, AD 1532, reproduced in Walsh, 1888: 83). According to Garrett (2006: 7) two sons of Nicholas St Lawrence lived successively in the house after their father's death. Meanwhile, the various lords of Howth sometimes used the lands at Killester directly and more often leased it to other individuals. For example, the record of a 1593 lease survives, whereby Nicholas St Lawrence, ninth baron Howth, leased 120 acres of arable, meadow and pasture at Killester ('the woods and underwoods excepted') for sixty-one years to Patrick Tailor of 'Cowlock' (Coolock) for the sum of £60 English money paid and £40 to be paid by Easter 1594 (NAI Lodge/4/449).

Not long after the lands of Killester were leased to Patrick Tailor in the 1590s, Ireland was plunged into a protracted period of unrest. These were to be some of the most turbulent years of Ireland's history, commencing with the Nine Years War from 1594 to 1603, increased intolerance of religions other than the established Anglican church of Ireland and a further eleven years of war beginning with the 1641 Rebellion and including the Cromwellian reconquest of Ireland. Whether or not the fields of Killester were directly impacted, one of the legacies of this unsettled era was a detailed documentation of the lands of Ireland, largely for the purposes of confiscating lands from Catholics and those who had been involved in the 1641 Rebellion and redistributing it to new British settlers and to pay creditors who had helped to fund the war effort. Although Killester remained in the hands of the St Lawrence family of Howth, these records provide a useful picture of the area in the mid-seventeenth century.

The Civil Survey (1654–6) recorded the ownership of land in Ireland as it had been prior to the 1641 Rebellion, with the proprietors identified by religion. The lord baron of Houth (Howth) was listed as owner of the 186-acre parish of Killester, valued 'by the jury' in 1640 at 40 pounds and 'by us' at 90 pounds. The survey gives an indication of the use to which the land was being put at that time. The majority, 140 acres, was in arable use, while there were forty acres of meadow and a further six of pasture. There was 'one faire stone house slated with several houses of office and a stone bawne valued at 300 pounds'. It can be conjectured that the stone house with bawn (a defensive wall) was on the site of the original motte and bailey. The document also mentions the walls of a decayed chapel and two or three thatched cottages. Although the parish tithes belonged to Christ Church, at this time they were in lease to Viscount Moore. None of the land had been redistributed from former 'papist' proprietors.

The Down Survey largely complemented the Civil Survey, involving the compilation of over 2,000 parish and barony maps outlining the location of the lands (held by 'Irish papists') likely to be forfeited. Because Killester was already in the ownership of a Protestant family, there was less detail recorded on the Down Survey than would be the case for forfeited lands. However, 'Killaster' is named in the Down Survey map of Co. Dublin. In the year 1659 the population of Killester was recorded as 13 'English' and 20 'Irish'. By contrast, the total population according to the 1881 census was 443 (Walsh, 1888, Appendix VI).

The evolution of Killester from earliest times

THE KILLESTER DEMESNE HOUSE

As peace was restored, many landlords across Ireland began to invest in their properties. In Killester, the slated stone house with its bawn that was described in the Civil Survey is thought to have been replaced by a new dwelling in the early eighteenth-century (Howlett, 1979: 71). Architectural historian Maurice Craig (1977: 166) described it as 'an unusually beautiful small early eighteenth-century single-storey house with a timber cornice, a steep roof with dormers, a central pediment and Venetian windows in the salient end-pavillions'. Neither the name of the architect nor of the client are known for certain. Rolf Loeber (1981) attributed the design of the new manor house to Thomas Burgh (1670 to 1730), stating that it was built for his colleague Chidley Coote (Loeber, 1981: 39). By contrast, Arthur Garrett (2006) claims that it was built by the St Lawrence family. Both suggestions have some merit.

We know that the Coote family had an association with Killester from the seventeenth century, presumably leasing it from the earls of Howth. D'Alton (1838: 123) reported that 'an inquisition of 1621 finds that Nicholas, baron of Howth, died in 1606 seised of Killester, three messuages, and seventy-two acres, &c. Chidley Coote, second son of Sir Charles Coote, the republican general, subsequently resided and died here, and bequeathed the estate of Killester to increase the jointure of his wife Anne.

4 H.G. Leask watercolour of the Killester mansion house, *c*.1907. [AD2477. NLI]

His interest was, however, derived from that of the Lord of Howth, in whose descendants the fee still continues.'

Colonel Chidley Coote MP, husband of Ann(e) or Alice Philipps and second son of Sir Charles Coote, the first Baronet Coote of Castle Cuffe, is thought to have been born *c.*1608 at Killester and died in 1668 in Limerick. Like his father he was a military man, and is described as 'Chidley, of Killester' in *Burke's Peerage*. Is it possible that he was in possession of the lands and an older house at Killester, which was then rebuilt by his son, also Chidley Coote (b. 1644, d. 1703)? There is a record of a 1681 judgment that divided the property of Chidley Coote, deceased, of Killester, Co. Dublin and Mount Coote, Co. Limerick, between his two eldest sons Chidley Coote and Sir Philips Coote. Perhaps, following this judgment, the younger Chidley Coote was now in a position to undertake the construction work.

Killester Demesne house was not the only grand residence to be constructed in the locality in the early eighteenth century. In the 1730s, Furrypark was built as a weekend villa for Quaker banker Joseph Fade, after whom Fade Street in Dublin city was named (Cannon, 1985). This simple but substantial house was built on land acquired from the Vernon estate. It has a storied history in its own right and is one of the earliest surviving buildings in the area. Further north towards Raheny, Sybil Hill was also constructed in the same decade. Lennon (2018) has explained how, by the late eighteenth century, there were more than twenty mansions with estates of several acres in Clontarf. Their occupants, members of the parliament and judiciary, bankers and merchants, were attracted to this rural location in close proximity to the city. A similar profile of affluent residents could be found in neighbouring Killester. These monied households would have contributed to the local economy, providing employment to domestic servants and outdoor labourers. Unfortunately, as tends to be the case, there is almost no trace of the everyday residents of Killester in the historical record.

KILLESTER FROM THE MID-EIGHTEENTH CENTURY

John Rocque's 1757 survey of the city, bay, and environs of Dublin is the first to show Killester in detail. Killester is depicted as a small village centred around what is now the intersection of Killester Avenue, Middle Third, and La Vista Avenue. There are ornamental gardens to the east and west of the Killester manor house, with a tree-lined avenue visible in the location of present-day Nun's Walk. The medieval church and churchyard are also depicted, with buildings constructed partially into the wall on the north-east and west sides. This map also shows that field boundaries just outside what became the nineteenth-century demesne boundary appear to radiate outwards, possibly indicating an older field system centred around what became the demesne lands.

5 The environs of Killester. Extract from John Rocque, *A survey of the city, harbour, bay and environs of Dublin … 1757*. [PC]. The demesne is highlighted.

One feature on Rocque's map of interest to archaeologists is the curvilinear feature which can be seen in an open field directly south of Killester House. This appears to be a depiction of a ditch and bank running eastwards from the field boundary, curving northwards to terminate in the boundary separating the field from part of Killester House's gardens. This could possibly indicate a much earlier boundary and may be a portion of the earthwork described by Father Kenny in 1934. If this is the case, and given that the earthwork as depicted on this map appears to respect Killester House, it is possible the house was constructed in the same location as an earlier fortification, with the earthwork depicted by Rocque forming an outer defence (Archaeology Plan, 2023). This cartographic evidence suggests that there may have been a continuity in the site of the mansion house at Killester with its predecessor from the medieval period. A possible further piece of evidence for this continuity, as previously noted, is the description of Killester in the Civil Survey of 1656 which refers to a 'bawne' wall, indicating the presence of fortified walls which could be remnants from the time of Brun (Simington, 1945: 173).

Although knowledge of the origins of the Killester demesne house is limited, information is available about some of its former residents. In common with many similar properties, the house and lands at Killester changed hands with some frequency. For example, a newspaper advertisement in 1777 offered a lease on 'the mansion house and demesne of Killester' where James Dunn esq. formerly lived, amounting to 23 acres 2 roods and 24 perches, and of the holding where Mrs Murray was living, comprising a dwelling-house and offices on 6 acres 2 roods and 11 perches. Proposals were to be made to the earl of Howth, who was the proprietor. These same plots are mentioned in the documentation for the Landed Estates Court, discussed below.

It seems likely that this 1777 advertisement marks the start of the occupation of the Killester property by Sir William Gleadowe Newcomen, as although there is a lease dated 26 June 1794, it seems that he was already in residence and had made improvements to the grounds by that date. Certainly, he is named on Taylor and Skinner's map, surveyed in 1777 (see Figure 6), while Sherrard's 1775 estate map of Richard Cook's property in the Longfield collection also shows his land as bordering that of W.G. Newcomen (see Figure 7). In 1779 the beautiful villa known as Killester Hall, previously residence of Richard Cook esq. (deceased), was being sold by auction by the sheriff of Dublin, due to unpaid debts. Cook had perhaps overextended himself, having spent upwards of 500 pounds on recent improvements to the property. The house, gardens and 17 acres and 2 roods of land were subject to the 'small' annual rent of ninety pounds (*Saunders' Newsletter*, 23 July 1779).

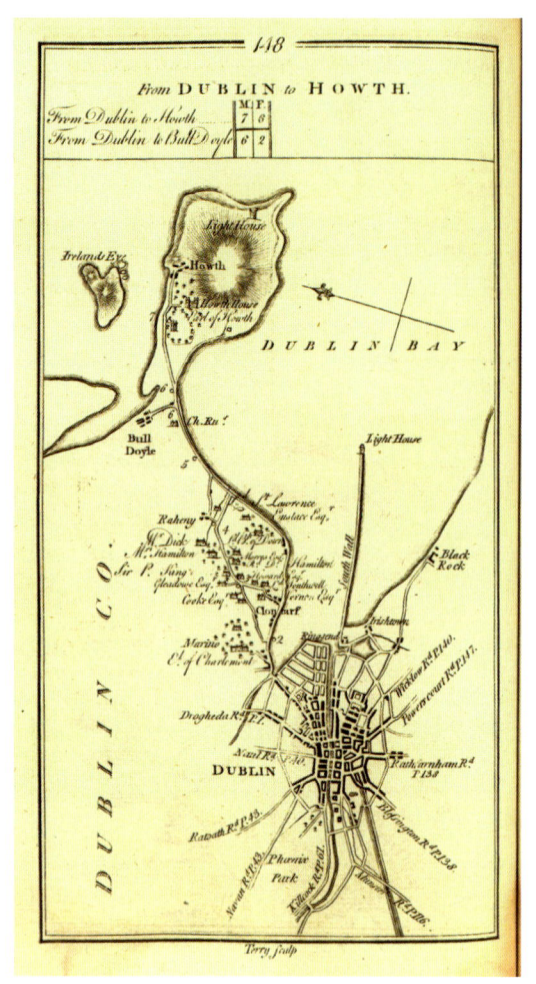

6 The road from Dublin to Howth. Map 148, depicting residence of 'Gleadowe Esq.' William Gleadowe Newcomen, esquire, is included in the list of subscribers. George Taylor and Andrew Skinner, *Taylor and Skinner's Maps of the Roads of Ireland, 1777*. [TCD]

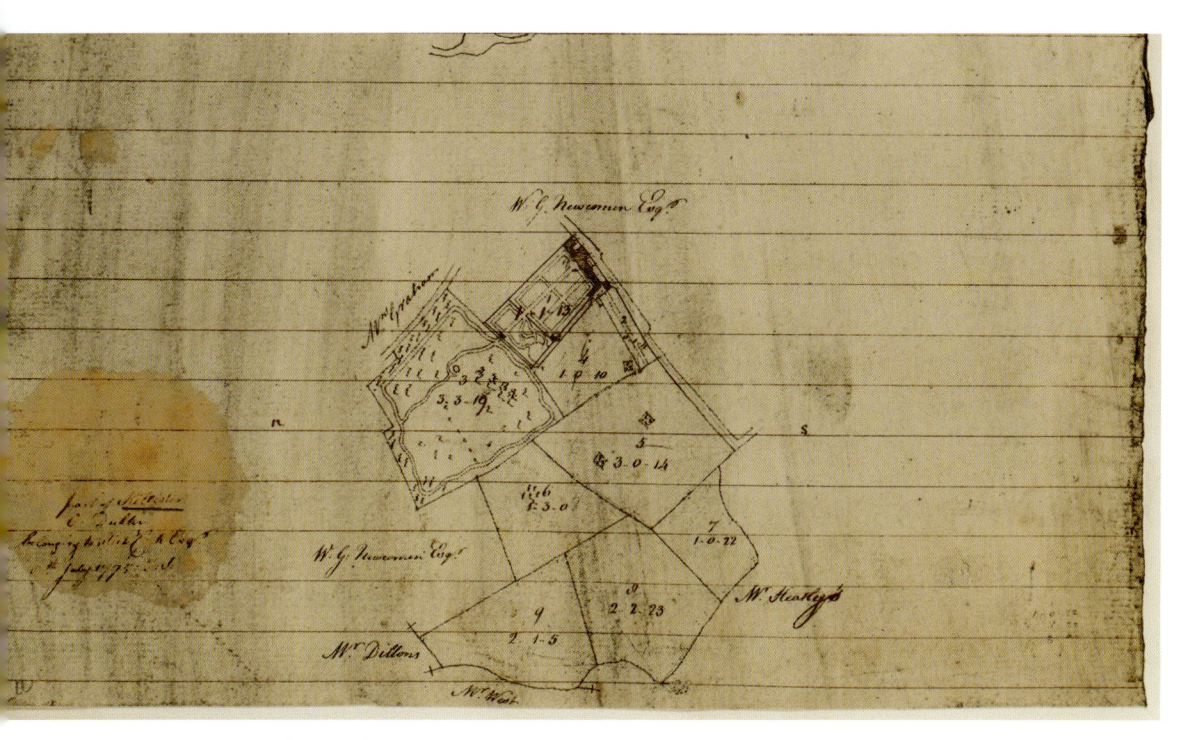

The evolution of Killester from earliest times

7 [A survey of] Part of Killester, Co. Dublin belonging to Richd. Cook Esq. 8th July 1775 [by] Thomas Sherrard, showing formal gardens to the rear of his residence and naming neighbouring landowners including W.G. Newcomen Esq., Mr Graham, Mr Dillon and Mr Heatly. [Longfield Collection MS 21 F. 51 / (118). NLI]

In the same decade, properties previously occupied by Robert Burton and by Mr Heatley in the neighbourhood of Killester were advertised, giving a flavour of the nature of the larger dwellings in the area at this time. The interest in the lease of Burton's 'good dwelling house' was being offered together with a coach-house, stabling for six horses and other offices, a large garden with eleven-foot-high brick walls and enclosing choice fruit trees, a large and well-chosen collection of flowers as well as a large asparagus bed. Two adjoining fields amounted to eight acres of choice meadow which had benefited from the application of 4,000 loads of manure in the previous three years (*Saunders' Newsletter*, 17 March 1773). A very highly manured meadow was also one of the attractions of Mr Heatley's concerns near Killester when these were advertised in 1776 (*Saunders' Newsletter*, 22 April 1776). A 'delightfully situated' property, the house was described as large and well built, fit for the immediate reception of a large family, while the large, well-laid out garden was planted with 'all kinds of choice fruit trees, and cropped with a variety of vegetables'.

The attractions of Killester to the prospective well-to-do resident were evident. As the advertisement for Burton's property remarked: 'It commands a beautiful and extensive prospect of the Bay, Wicklow mountains etc and very convenient for bathing, being but a small mile from Clontarf'. Similarly, when the interest in the late Owen McDermott's lease on a property in Killester was being sold in 1784, it was noted that the location was within two miles of Dublin (although a subsequent advertisement specified three miles!), while the house itself (Killester Lodge) was large and convenient, the ground highly manured, the garden large and well cropped. There were 17½ acres of land 'laid out to great advantage'. The entire property was subject to a rent of £90 per annum and the interest of the lease had previously been acquired in 1781 for a fine of £200 and £50 for the fixtures of the house (*Saunders' Newsletter*, 12 March 1784; *Saunders' Newsletter*, 7 April 1789). Another well-known resident of the area was Sir John Blaquiere, who served as chief secretary for Ireland between 1772 and 1776. He occupied Jane Park on ten acres prior to April 1789.

THE GLEADOWE NEWCOMEN FAMILY

The Gleadowe Newcomen family occupied the Killester mansion house for two generations, beginning with William. William Gleadowe, born *c.*1741, was the son of banker Thomas and his wife Teresa, of Castle Street in Dublin, and became a partner in his father's private bank. After marrying Charlotte Newcomen, only child and heiress of Edward Newcomen's estate at Carrigglas (sometimes spelled Carrickglass), Co. Longford (worth £5,000 per annum), on 17 October 1772, he assumed her additional surname and coat of arms (Garrett, 2006: 11). On 9 October 1781 William was created a baronet, Carrickglass in the baronetage of Ireland. Between 1790 and 1800 Gleadowe Newcomen was the member of parliament for Co. Longford in the Irish House of Commons. Originally opposed to the Act of Union, having subsequently voted in favour he was well rewarded. One of the benefits included the creation of his wife Charlotte as Baroness Newcomen, with the remainder to his male heirs. Following the Acts of Union 1800, he represented Longford in the House of Commons of the United Kingdom between 1801 and 1802, making way for his son Thomas at the 1802 general election. Newcomen was made a Viscount in 1803, with his wife becoming Viscountess Newcomen, also in her own right. Richard Lovell Edgeworth, father of novelist Maria, had voted against the Union, and summed up his opinion of Sir William in the following jingle as quoted by Garrett (2006: 12):

> *With a name that is borrowed, a title that is bought*
> *Sir William would fain be a gentleman thought;*
> *His wit is but cunning, his courage but vapour,*
> *His pride is but money, his money but paper …*

The evolution of Killester from earliest times

8 Portraits of the Gleadowe Newcomen family. Benjamin Wilson (1721–88), *Portrait of Sir William Gleadowe Newcomen seated at a table and the Hon. Thomas Newcomen as a child*, oil on canvas, 73.5cm x 61cm [TA].

Thomas Hickey, *Charlotte Newcomen, Lady Gleadowe Newcomen, 1st Viscountess Newcomen (c.1747–1817) with her daughters Jane, Teresa and Charlotte in a garden*, oil on canvas, c.1767–80. [NT]

NEWCOMEN'S BANK, ARCHITECTURE, TASTE AND PATRONAGE OF THE ARTS

Whatever the received wisdom about William Gleadowe Newcomen and his social climbing, there is no doubt that William became hugely influential not just as a banker but also as a patron of art and architecture of the most sophisticated kind. The role of private banks such as Newcomen's was somewhat different to that of modern banking, which emerged in Ireland following the Banking Act of 1821. During the eighteenth century, a wave of unprecedented prosperity in Dublin had led to the establishment of private banks that provided short-term credit and facilitated the movement of clients' money from one location to another. However, a number of banking failures occurred in the mid-eighteenth century, leading to the failure of most of the Irish private banks, with just two surviving untarnished – the La Touche Bank and Gleadowe's (later Newcomen's) bank (O'Neill, 2011). The importance of these two banks was reflected in their prominent location on Castle Street, close to the seat of power at Dublin Castle. Both were impressive structures, as befitted the status and influence of those conducting business there. The La Touche Bank has since been demolished, leaving Newcomen's bank as a unique survival in Dublin, which is considered one of the finest eighteenth-century buildings in the city.

In about 1778 William Gleadowe Newcomen employed architect Thomas Ivory, one of the significant figures in the construction of Georgian Dublin, to design his new premises at Castle Street. Incidentally, records show that James Hoban, the Irish architect who went on to design the White House, worked for Ivory on the plans. The site was exceptional, with its side elevation facing the recently completed Royal Exchange (now City Hall) at the entrance to Dublin Castle. This ensemble of historic buildings of the highest architectural quality has been described as one of the signature set-pieces of the city, which has been portrayed in numerous artistic representations from the eighteenth century onward.

Newcomen's Bank, completed in 1781, was designed to be both a private banking house and a private residence. Casey (2005: 365) describes the building as an 'enigmatic and exquisitely made building, the sole instance when Ivory's built work matches up to his spectacular drawing skills'. The exterior of the building used sharply detailed Portland stone, the material reserved for the best public buildings in the Georgian city. Ivory's original building comprises a three-room plan with an impressive central top-lit open-well stair hall including high quality decorative plasterwork. The rooms were elegant and well proportioned. Select clients would have been welcomed to the bank parlour on the first floor (O'Neill, 2011). This was an impressive oval room featuring a *trompe l'œil* ceiling attributed to the Italian painter Vincent Waldré, whose other works include the ceiling of St Patrick's Hall in Dublin Castle. This building, still usually referred to as Newcomen's Bank, was bought by the Hibernian

Bank after 1825 and subsequently extended. It later became the Rates Office and is in the ownership of Dublin City Council.

William Gleadowe Newcomen continued his pattern of employing the best architects and artisans of the era when considering plans for his country estate. Having gained control of Carrigglas Manor in Co. Longford through his marriage to Charlotte Newcomen, he employed James Gandon to design a new house and associated estate buildings. However, only the double courtyard, stable and farmyard complex and main entrance gates were completed by James Gandon. Carrigglas Manor was leased to, and later bought, by the lawyer Thomas Lefroy, in the early nineteenth century and it was he who built the present Tudor-revival Carrigglas Manor (*c.*1837).

Newcomen was elected company secretary of the Royal Canal Company at its inaugural meeting on 13 November 1789 and acted as treasurer for the company. His involvement in the enterprise is recalled in the naming of Newcomen Bridge, which spans the Royal Canal at North Strand, and was completed in 1793. In 1804 he bequeathed a Royal Canal debenture for £100 to Drumcondra parish, the interest of which was intended to be used to purchase bread and provisions for the poor of the parish (Garrett, 1970: 61).

THE NEWCOMEN INFLUENCE AT KILLESTER

Despite his busy commercial life, William Gleadowe Newcomen found time to undertake improvements at his Killester estate. In 1794 builder William Waldron was employed to build a new greenhouse, peach-house, grape-house, as well as portico, ceiling cornice, Gothic mouldings to windows and Gothic window stools (DIA.ie). The *Dictionary of Irish architects* also mentions that John Sutherland (1745–1826), the most celebrated landscape architect of his day, advised on a park at Killester, possibly in the 1790s. It is quite likely that this was in relation to improvements being undertaken at the Killester Demesne and would be in keeping with the other activities and improvements by Newcomen at this time. Indeed, the documents of the Landed Estates Court (discussed later) refer to the 1794 lease whereby the earl of Howth demised the mansion house of Killester and twenty-seven acres of land to Sir William Gleadowe Newcomen, baronet. A memorandum was included to the effect that Newcomen, 'having lately planted a considerable number of timber and other trees on said premises, his Executors, Administrators, and Assigns, were to have full liberty to cut down, dig up, and remove not only the trees already planted by him, but such as he should thereafter plant.'

The names of the fields belonging to the house give an indication of their nature and uses: 'the Cock Park, Hen Park, Sycamore Park, Horse Park, Grove, and the Gardens', amounting to 27 acres 1 rood and 39 perches. The property leased to Sir William Gleadowe Newcomen by the earl of Howth also included a further 23 acres

9 View of the entrance to Killester Demesne by Edward McFarland, 1853. This was the entrance identifiable on the first edition Ordnance Survey plan where the Howth Road curved to the east. [PD 1986 TX 8. NLI]

1 rood and 18 perches in Irish plantation measure, which included a dwelling-house, fields called the Little Wood Park, the Deer Park, the Coal Fields, the Three Parks, and the Well Meadow, with a cabin and gardens which had formerly been Hannah Corker's. Additionally, a sliver of land bounded to the north by the road from Dublin to Howth included a dwelling-house and ground occupied by four thatched cabins with their gardens. These very humble abodes are rarely identifiable in the historical record but become evident again in Griffith's Valuation and in the census returns of the early twentieth century.

Killester Demesne was accessed through an elegant gateway with two built-in lodges, which stood opposite to where the present-day shops are located in Killester village. This is depicted in McFarland's 1853 illustration (see Figure 9) but was already in existence at the time of the first edition Ordnance Survey, and probably dates to the 1790s. This elaborate gate screen was designed by James Gandon, one of the most celebrated architects working in Ireland in the late eighteenth and early nineteenth centuries. As noted above, Gandon undertook a series of designs at Carrigglas Manor in Co. Longford for its owner, Sir William Gleadowe Newcomen, c.1795. Gandon's surviving drawings for the gateway in the National Library of Ireland do not match the completed entrance to Carrigglas, however, and appear to be those for Killester.[5] Its high-quality design reflects the importance of Killester Demesne and the desire to represent this status to those entering the estate or simply passing by on the Howth Road. A winding avenue led from this formal entrance through parkland studded with trees to the mansion which stood on a slightly elevated site. A shorter and less formal access is depicted on the first edition Ordnance Survey map leading from a gate lodge at a sharp bend in Killester Avenue (close to the modern-day junction with La Vista Avenue). The main pair of gate lodges on the Howth Road appear to have survived into the twentieth century. The central gate piers were taken and re-erected at one of the western entrances to the Howth Castle estate, off Carrickbrack Road, where they remain.

By the early nineteenth century, the proximity of Killester to the fashionable delights of sea-bathing at Clontarf were increasing the attraction of houses in the area. These were now being let, furnished, for the season (see, for example, *Saunders' Newsletter*, 25 May 1802). The advertisers of 'Newhall, Killester' were appealing to a variety of markets when, in 1803, they offered the dwelling-house, with offices, gardens and thirteen acres of ground, either for the season or for a long term of years. They were keen to note that it adjoined Sir William Gleadowe Newcomen's demesne, probably hoping that this would confer a certain cachet. The Newhall property offered 'the best meadow in Ireland, subdivided into six fields, with piers, gates, and full-grown hedges', while the house consisted of a parlour and drawing room, as well as six bedrooms (*Saunders' Newsletter*, 13 July 1803).

In 1801, Archer (p. 92) described Killester as a 'pleasant village' and provided a glowing description of Killester, the seat of Sir William Gleadowe Newcomen, as follows (p. 97): '... with a spacious house, a demesne of near forty acres, well wooded; the walks are judiciously laid out, so as to form a compleat country residence, though situated within a mile and a half of the capital. The gardens are elegantly disposed, a large extent of glass well furnished with pines [pineapples], grapes, etc., of the first flavour'. A briefer, but similarly positive, description of Killester was published almost a quarter of a century later in George Newenham Wright's *Historical guide to the city of Dublin* (1825: 245) where it was described as 'a beautiful demesne of about 50 acres, with an excellent house. In the garden are graperies and pineries of great extent'.

Upon William Gleadowe Newcomen's death at Killester in 1807 (the viscountess died a decade later at Bath and is interred there) he was succeeded in business by his son, Thomas Gleadowe Newcomen. Thomas had been educated at Eton and Oxford, was admitted to the Bar at Lincoln's Inn in 1794 and was MP for Longford from 1802 to 1806. However, William was already significantly in debt to his own bank at the

10 The Newcomen Bank at Castle Street. [DCC]

time of his death. His son Thomas Viscount Newcomen followed his father's example, borrowing extensively. In an 1892 memoir, William John Fitzpatrick described how: 'For years he lived alone in the bank, gloating, it was wildly whispered, over ingots of treasure, with no lamp to guide him but the luminous diamonds which had been left for safe keeping in his hands. Moore would have compared him to "the gloomy gnome" that dwells in the dark gold mine. Wrapped in a sullen misanthropy, he was sometimes seen emerging at twilight from his iron clamped abode'. (McGrath, 2012). Garrett (2006: 12) records a similar tale, noting that 'wealth and titles brought little happiness to this abject man. His life was sordid and dissolute, and he has been accepted as being a most unpleasant man.'

Whatever the truth behind these accounts, Newcomen's bank was clearly mismanaged. A number of private bank failures were experienced in Ireland between the late 1790s and the early 1820s. Newcomen's bank collapsed in 1825 and the building was subsequently acquired by the Hibernian Bank. The banking failure ruined his own family and many clients. In the face of this scandal, Thomas Newcomen shot himself. He was 48 years of age. Although the *Morning Post* attempted to allay rumours by reporting that the coroner had found no marks of violence on his body, this would ultimately prove not to be the case (*Morning Post*,

11 The ceiling in Newcomen Bank. [DCC]

12 Gleadowe Newcomen tomb at Drumcondra churchyard. Photograph: Ruth McManus.

22 January 1825). As the *Dublin Evening Post* reported on 18 January 1825, the announcement of Newcomen's death, combined with the fact that the doors of the bank on Castle Street had been closed, 'caused a great sensation in town during the whole of yesterday'. Within a few days, a meeting of creditors discovered that the bank had been in difficulties, perhaps even insolvent, for the previous twenty-five years (*Dublin Evening Post*, 25 January 1825). Following his death in 1825, all his titles became extinct, and his estate was ordered to be sold in the Court of Chancery. The Annual Register of 1825 reported that 'the whole of the unsettled estates are subject to the debts of the house'. Initially it seemed that his family would be left in penury, but by June it was being reported that the estates would likely yield more than initially anticipated, so that Newcomen's illegitimate children would benefit from a surplus (although he never married, he had a long-term relationship with Harriet Holland who bore him eight children) (*Drogheda Journal*, or *Meath and Louth Advertiser*, 23 July 1825). Thomas Newcomen was interred in the family tomb at Drumcondra (see Figure 12).

In 1831 the Chancery Court ordered 'the Mansion House and Demesne Lands of Killester, the late residence of the late Viscount Newcomen, and the adjoining lands at Killester and Clontarf set to tenants' be sold to the highest and best bidder. The household contents were also auctioned off. The description of the items being sold gives a good indication of the degree of luxury and lavish lifestyle which had been enjoyed by the residents of Killester House:

> a quantity of Household furniture, the Property of the late Lord Viscount comprising mahogany parlour chairs; cane seat drawing-room ditto; Grecian and Indulging sofas; dinner, breakfast, card, sofa, and work tables; pier-glass and cabinet; carpets and rugs; dumb waiters; plate buckets; tea stores; forty-two fine engravings by Piranesi; curtains and draperies; brass fenders and fire irons; hall chairs; an eight-day clock; a richly carved and gilt picture frame.
>
> Also, the entire excellent Furniture of eleven bedrooms; in which are four-post, elliptic, waggon-roof, and other bedsteads and curtains, with the best feather beds, curled hair mattresses and bedding; a number of mahogany wardrobes, and several cases of drawers; dressing tables and basin stands, with marble tops; bidets and commodes; toilet glasses; cheval ditto; bed-room chairs – carpets, house linen, table cloths and napkins; china, and glass; a variety of kitchen and dairy requisites; a large quantity copper culinary articles; deal presses and tables; mangle.
>
> The entire utensils of a brew-house, complete; a bathing box on wheels; two slipper baths, and a few greenhouse-plants, with numerous other articles.
>
> (*Dublin Evening Mail*, 15 July 1831).

KILLESTER AFTER THE NEWCOMENS

Favourable descriptions of Killester continued to appear in publications in the 1820s and 1830s. Brewer's 1825 *The beauties of Ireland* described Killester as 'an agreeable village, adorned with one extensive demesne, and several handsome abodes of inferior extent' (Brewer, 1825: 231). He described Killester House, 'the principal residence in this village,' as 'a mansion of extensive proportions and pleasing character, surrounded by a demesne laid out with much correctness of taste' (Brewer, 1825: 232). By the 1830s, the parish of Killester, which comprised 228 acres, had just 113 residents, of whom eighty-six were Catholics (Lewis, 1837; D'Alton, 1838). Killester House, now the property of General Luscombe, was the most notable residence in the parish. D'Alton (1838) notes: 'its gardens were formerly much admired. The hall is spacious, the reception-rooms good, and the demesne tastily laid out, displaying fine vistas of the bay and its southern shores, some winding wooded walks, and one straight arcade still termed "the nuns' walk".' Lewis' *Topographical dictionary* (1837) also included Killester House in its list of important residences. Interestingly, it states that the house 'incorporated the remains of an old abbey, and in the demesne is a fine grove of lime trees.' Gwynn and Hadcock's (1970) listing of Killester, Dublin, note the ruined church, and further refer to Cassell's gazetteer, which suggested that some remains of a monastery existed in Killester House. For the unclassified monasteries covered in their volume, including Killester, they suggest that 'many were probably not religious houses in the true sense, being churches or chapels, granges or domestic buildings, the sites or ruins being embellished with the name "abbey" etc.'. If that is the case, then it is possible that there were remains of manorial buildings at the site of Killester House which were subsequently labelled with the term 'abbey', which is later incorporated into the placenames of 'Abbeyfield' and 'nun's walk'.

From the 1820s the first surviving national list of occupiers of land was compiled for the purpose of assessing tithes, the tax payable by occupiers of agricultural land (irrespective of religious affiliation) which formed the main source of income for the parish clergy of the Church of Ireland. The Tithe Applotment Books provide the names of occupiers of land, the size of their holdings (in Irish acres, i.e., 1.62 times larger than statute acres) and the tithe payable. The records in the Tithe Applotment Books of 1834 list twenty-two individual occupiers of the 280 statute acres in Killester parish. Of these, the most substantial holding was that of Commissary General Luscombe (43a. 1r. 4p.), with other significant holdings by John Smith (28a. 2r. 13p.), Representatives of Henry Cooper (20a. 1r. 24p.), Michael Cox (19a. 3r. 29p.), George Farran (19a. 3r. 29p.), Edward Byrne (17a. 36p.), Captain Philip Fitzpatrick (16a. 3r. 30p.), Nathaniel Low (15a. 28p.), T. Richardson (14a. 3r. 28p.), George Wilson (13a. 2r. 39p.) and George Symes (10a. 29p.). Relatively smaller holdings were recorded for Mrs Hargrave, Edward Murphy, Matthew Mooney, J. Bingham, Mr Irwin, William

Alley, James Chambers, Thomas Carolan, Richard Peters and the house and gardens of Lady Fane. An untenanted plot was listed as formerly having been part of Hickey's Nursery. This was presumably the business of P.M. Hickey & Company, nursery and seedsmen, who had premises at Sackville Street and a nursery at Clontarf Strand from at least 1817 (*Dublin Evening Post*, 1 February 1817).

One of the individuals named in relation to the tithes, Thomas Carolan, was a shopkeeper as well as a farmer. By 1855 he held the lease of Hollybrook, a coaching inn on the Howth Road which was first licensed in 1798 and which stood on the site of the present-day Harry Byrne's pub (Gogarty, 2013: 84–5). The importance of the Howth Road had increased from 1818 with the designation of Howth as a mail packet station. This was now the main routeway for the postal service between Dublin and London (until it was switched to Kingstown/Dún Laoghaire in 1834). Thomas Telford was commissioned to design the 'London to Dublin' road via Holyhead and Howth. One of his distinctive milestones can still be seen on the Howth Road at Killester (Rickard, 2017: 20). The inn at Hollybrook was one of just two mail despatch depots on the route from Dublin, with the other being in Raheny. There were livery stables, a coach yard and a farrier's yard, so that the horses could be tended to, rested or exchanged for the next stage of the journey.

Some of the names found in the records for the tithes also appear among the list of main houses noted by Lewis in 1837. In addition to Luscombe's property at Killester he listed Maryville, the seat of A. Barlow, Esq.; Woodville, of J. Bingham, Esq.; Hollybrook House, of W. McDougall, Esq.; Hollybrook Park, of G. Symes, Esq., a wine merchant (Gogarty, 2013: 84); Killester Lodge, of G. Wilson, Esq.; Clontarf Strand, of J. Chambers, Esq.; and Oatley, of G. Farran, Esq. Some of these residences might today be more strongly associated with Clontarf, while others are omitted in this entry and instead appear in the listings for Artane and Raheny. Edward Murphy's name appeared in the Killester entry of Pettigrew & Oulton's *Dublin almanac* for 1837, where he was described as a 'rural improver'. The other residents listed for Killester were General Luscombe, George Wilson of Killester Lodge, William Nugent of Killester Abbey, and Philip Fitzpatrick. By the 1847 edition, Luscombe's second residence at 96 Leeson Street was also included in the listing, while Mr Wilson had been replaced at Killester Lodge by Hugh Morrison, whose town address was 27 Castle Street. Fitzpatrick was gone and a Miss Downes was living at an unnamed residence in Killester.

THE FIRST ORDNANCE SURVEY MAP OF KILLESTER

The first map to show the full area of Killester is the first edition six-inch (1:10,560) Ordnance Survey map of the locality, surveyed in 1836 and published in 1843. It

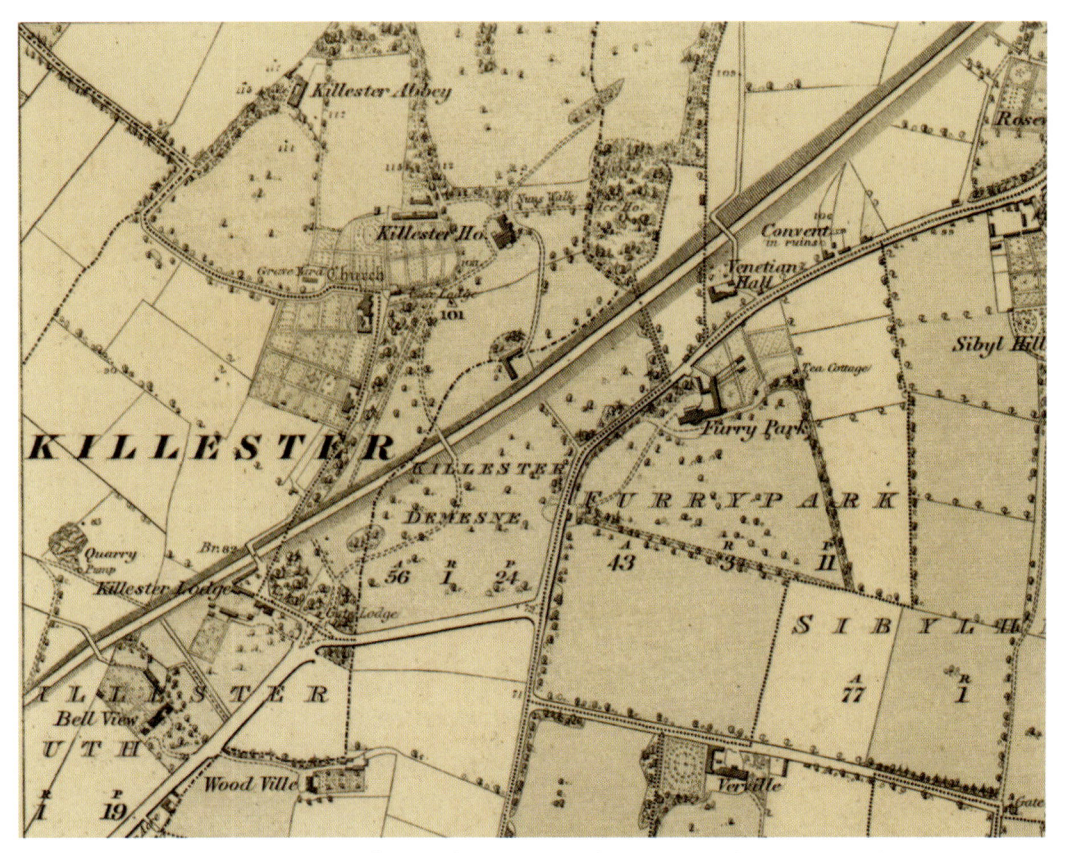

The evolution of Killester from earliest times

13 Killester House. Extract from Ordnance Survey plan, 1:10,560, sheet 19, 1837 edition.

shows the important houses listed above. Following the practice used across the country, the demesne was shaded on the map and its perimeter demarcated by trees. The gardens appear to have extended further north than depicted on Rocque's 1757 map. The field immediately north of the mansion house had been turned into parkland bordered by thick trees on the north, west, and part of the east sides. To the west of the House, buildings present on Rocque's map no longer appeared and two long buildings, possibly stables, had been constructed in the area of what is now Abbeyfield Lawn. To the east, joined to the house by Nun's Walk, was a wooded area containing an ice house.

Interestingly, two named houses were depicted within the demesne. In addition to the mansion known as Killester House, a second dwelling known as Killester Abbey was located to the north-west. Slightly surprisingly, although a townland called 'Killester Demesne' was identified on the map, the majority of the lands of the actual house and demesne were not located within this townland, but rather in 'Killester North'. Another feature of the first edition sheet is the 'convent in ruins' located slightly to the east of Venetian Hall, which perhaps indicates some historical evidence for the local belief in a historic 'convent' or 'abbey' or monastic presence in the locality. Nevertheless, this site is significantly removed from the long-standing cluster of church and graveyard that form the core of Killester. A quarry was located in the lands of Killester South, close to the new railway. Construction of the Dublin and Drogheda railway began in October 1840, with the official opening of the line taking place, including a ceremony at Raheny, in May 1844. The line cut through the demesne lands to the south of the house, with some parkland and the gate lodge lying on the south side of the railway track.

As noted above, one of the dwellings on the map was named Killester Abbey. It seems quite likely that the immediate origin of the present-day street name Abbeyfield was the house known as Killester Abbey, which was located immediately to the west of the present-day street (Killester Park), and which survived after the houses were built. It is quite possible that the house known as Killester Abbey did not have any specific association with a historic monastic site, but rather reflected its location not far from a ruined church and the need to distinguish this dwelling from others in the vicinity, namely Killester Demesne, Killester Gardens, Killester House and Killester Park, all of which are also included on the first edition Ordnance Survey.

It is not entirely clear when the Luscombe family came into occupation of the demesne, but it was possibly following the advertising of Killester House to let in June 1836 (*Saunders' Newsletter*, 1 June 1836). The advertisement gives an indication of the many qualities of the house and grounds, as follows:

> KILLESTER HOUSE, The Residence of the late Lord Newcomen.
> TO BE LET, on Lease. Furnished, Unfurnished, or the Interest to be Sold in this genteel residence, near the third milestone on the Howth road, a few minutes walk from the seashore; it is situate within a park-like fence, inclosing twenty-five or forty-four plantation acres and bounded by a belt of trees, with extensive walks and shrubberies. The House consists of a splendid Entrance Hall, with Dining, Drawing, and Breakfast-rooms, and ten Bed-rooms, with every accommodation for a family of distinction. The Garden, Graperies, Green-house, and Ice-houses are of the very first description; and there is also a compact Farm-yard and Offices of every kind. For particulars inquire of Mr Charles Parry, No. 2, Newcomen Terrace, North-strand.

The Luscombe family, like their predecessors the Gleadowe Newcomen family, were wealthy and influential. Thomas Popham Luscombe, born *c*.1780, married Catherine Tooke Robinson in London in 1821. He had been promoted to commissary general of the army in Dublin in 1826. They lived at Killester manor house, which they leased from Lord Howth, at least until Luscombe's death in 1855. According to Garrett (2006: 13), Luscombe maintained the estate to a very high standard. After his tenancy,

The evolution of Killester from earliest times

14 Portraits of Thomas Popham Luscombe and Catherine Tooke Robinson Luscombe upon their marriage, by Benjamin Delacour. [FA]

it was leased to his son Tooke, who was followed in residence by Lieutenant Arthur Lynch, a surgeon and physician with the Royal Army Medical Corps (see below).

Despite the claims that it was well kept, there is some contradictory evidence from O'Flanagan's (1837) transcription of the Ordnance Survey name books (see RIA archive). This suggests that Killester House was already experiencing a gradual decline, with the description noting a one storey, slated house, slated out-offices and a garden 'in bad order'. This was described as being the property of Commissioning General Larcombe 'who resides at Donnybrook and cannot let the house'. This is rather at odds with the other accounts of the period but is a first-hand account of the time.

MARKET GARDENING AT KILLESTER GARDENS

One of the dwellings depicted, but unnamed, on the first edition Ordnance Survey map was the house and grounds known as Killester Gardens. The premises was appropriately named, as it came to be used for market-gardening purposes. It appears to have been part of the lands in possession of Thomas Popham Luscombe, as he leased Killester Gardens, with 2 acres 1 rood and 37 perches, to Patrick Kirby for a thirty-year period from 1852. Notably, the terms included a special covenant to prevent certain trades or business being conducted there. This stipulation was perhaps made because Kirby was in business. The Kirby family continued their occupation of Killester Gardens until at least 1927 (by 1932, Thom's records a new occupier, T. Collen).

By 1853 Kirby's of Killester Gardens were advertising fresh arrivals of 'tomatas, Spanish onions, Kent filberts etc' to their Home and Foreign Fruit Warehouse at 13 Upper Sackville [O'Connell] Street (*Saunders' Newsletter*, 5 September 1853). Proprietor P. Kirby, fruiterer, continued to advertise his wares and an advertisement in advance of All Hallow's Eve suggests that he had been in business for the previous five years (from 1848) (*Saunders' Newsletter*, 25 October 1853). The enterprise was successful at the Royal Horticultural Society of Ireland May exhibition in 1860, where the second prize for mushrooms was awarded to Mr Kirby of Killester Gardens (gardener Mr Campbell) (*Irish Times*, 26 May 1860). By the mid-1870s, Kirby's was described as a florist as well as fruiterer, and had added premises at 9A Nassau Street (*Irish Times*, 23 December 1876). Advertisements continued to 1881 in the *Irish Times*. Early in the new century Kirby, now described as a florist, was advertising from 43 Mary Street (lately of Sackville Street), offering 'fresh flowers, fruit and vegetables daily from Killester Gardens' (*Freeman's Journal*, 4 September 1900, with latest example in 1907). Patrick Kirby, fruit merchant, died at Killester Gardens on 11 January 1903, aged 82 years,[6] followed by his widow Marcella, aged 86, in 1909.[7] The enterprise was an important employer in the locality. In August 1913, when there was a strike among farm hands in north County Dublin, it was noted that twenty men at P. Kirby's in Killester were on strike, as were a further twenty at T. Dunne's in Raheny and sixty-eight at F. Grogan's in Coolock (*Evening Herald*, 15 August 1913). Census returns from the beginning of the century list the six-roomed dwelling in Killester South and, in 1911, as being on Killester Lane. The 1901 census recorded 40-year-old market gardener Joseph Kirby and his two unmarried sisters, while in 1911 Joseph (now listed as being 46 years old, not 50 as might be expected) is described as a master gardener with his 42-year-old brother Bernard as assistant gardener.

KILLESTER IN GRIFFITH'S VALUATION

The Primary Valuation of Ireland, better known as Griffith's Valuation, was the first, full-scale valuation of property in Ireland since Petty's survey of 1658. The valuation for Dublin was completed in the years of the Great Famine and provides a snapshot of landownership and occupation at that time. It lists names of occupiers of land and buildings, the names of those from whom these were leased, and the amount and value of the property held. The records underlying the valuation have been digitized and made available by the National Archives. Those for Killester comprise 'house books'

The evolution of Killester from earliest times

15 Primary (Griffith) valuation for Killester Demesne – the summary and the surveyor's maps. [Ask About Ireland]

dating to 1845 and 1846 and record the occupier (householder), property name, description of the property and amount of valuation in money. In total, ninety-six individual listings cover Killester North, Killester South and Killester Demesne. These detailed handwritten entries provide a snapshot of the locality at a point in time, detailing all buildings including dwellings, barns, cow-sheds and privies, together with a classification based on their construction materials.

Within the fifty-six acre townland of Killester Demesne, Griffith's Valuation noted that General P. Luscombe was the main lessor, with the majority of the demesne under the immediate occupation of James Wynne Esquire. The mansion at Killester was valued at £48. The 'house book' further details the dimensions of each room within the Killester Demesne house, which is classified as 1C+ (i.e. 1 = a slated dwelling-house built with stone or brick and lime mortar, C+ = old but in good repair) and 4C+ (referring to the basement of a dwelling-house, old but in good repair). The surveyor included a note stating that 'the house is very old but kept in good sound order and repair'. A lengthy list of out-buildings, including slated sheds, an old disused tea room ('of no value'), a cow-shed, privy, barn ('in farm yard') and store rooms, gives an indication of some of the uses to which the extensive buildings on the property were being put. By now the railway had pierced through the land, as seen in the occupation of just over three acres by the Dublin and Drogheda Railway Company. It should be noted that Killester Demesne was located, somewhat confusingly, within the Clontarf parish. A more extensive area, the parish of Killester, included Killester North (including the 'village of Artaine') and Killester South.

The maps used for the valuation show the mansion house, labelled as 'Killester Demesne', with Killester Abbey to the north-west and to the south Killester Gardens and Killester House. Killester Park is located south of the railway line.

KILLESTER DEMESNE IN THE LANDED ESTATES COURT, 1863

Thomas Popham Luscombe of Killester House, former Commissary General, died on 15 March 1855 aged 73 and is buried at Mount Jerome cemetery. His sons William Hill Luscombe and Thomas Charles Popham Luscombe died in 1856 and 1858 respectively. This may have precipitated the ultimate sale of the Killester Demesne via the Landed Estates Court in 1863.

The lands at Killester Demesne which had been occupied by Thomas Luscombe, amounting to 117 acres 1 rood and 37½ perches, were sold at the Landed Estates Court on 12 June 1863. The papers associated with the sale provide details of the demesne and associated residences not just at the time of sale, but also in terms of earlier land transactions reaching back seventy-five years. The mansion and a little over fifty acres fell under a lease dated 25 June 1794 from the earl of Howth to Sir William Gleadowe Newcomen (discussed above). A second lease was dated 24 December 1788 whereby Bartholomew Levey demised part of Killester (part of the holding of the late Joseph Fade and comprising 6a. 2r. 7p.) including dwelling-house and offices to John

The evolution of Killester from earliest times

> **In the Landed Estates Court, Ireland.**
> # COUNTY OF DUBLIN.
>
> *In the Matter of the Estate of*
> Sir JAMES DOMBRAIN, and JOHN ROSS MAHON, and JOHN CASEMENT, Esquires,
> Trustees of the will of THOMAS POPHAM LUSCOMBE, Esq. Deceased.
> *Owners, and Petitioners.*
>
> ## Rental and Particulars of Sale
> OF THE
> ### MANSION HOUSE AND DEMESNE LANDS OF KILLESTER,
> Held under Lease for 91 Years from 1st May, 1794.
> **Other part of said Lands of KILLESTER,**
> Held under Lease for the respective terms of 82 Years from 29th September, 1788, and 14 Years from 1st May, 1871.
> **Other part of said lands of KILLESTER,**
> Held under Lease for 9 Years from the 29th September 1860, if the Lessor therein should so long live.
> **And other part of said Lands of KILLESTER,**
> Held under Lease for 9 Years from the 29th September 1860, if the Lessor therein should so long live. And
> ### Part of the lands of FURRYPARK,
> Now forming part of said Demesne lands of Killester, containing 1a. 0r. 18p. statute measure, held in fee.
> All situated in the Barony of COOLOCK and County of DUBLIN, and found to contain (by a survey thereof made in the year 1856 inclusive of said piece held in fee and exclusive of the portion occupied by the Line of Dublin and Drogheda Railway Company,) 117a. 1r. 37½p. statute measure,
> ### WHICH WILL BE SOLD
> IN ONE LOT, BEFORE JUDGE HARGREAVE,
> ### AT THE LANDED ESTATES COURT, INNS-QUAY, DUBLIN,
> *On Friday, the 12th day of JUNE, 1863, at the hour of 12 o'clock, Noon.*
>
> For Rentals and further particulars, apply at the Office of the Landed Estates Court, Four Courts, Inns-quay, Dublin; or to
> **HENRY OLDHAM,**
> **Solicitor having carriage of Sale,**
> ***No. 42, Fleet-st., Dublin.***
>
> Dublin:—Printed by W. KIRKWOOD, 33, Great Brunswick-street.

16 Extracts from Landed Estates Court records for Killester, 1863 showing the rental and particulars of sale and a map of the demesne and lands (on page 32). [Find My Past]

Dowdall. A reversionary lease dated 26 June 1794 saw the same premises demised by Thomas, then earl of Howth, to Sir William Gleadowe Newcomen. However, this was disputed by the then current earl of Howth and claimed to be invalid.

The details show that, while most of the estate was ultimately owned by the earl of Howth, the proprietor of Clontarf, John Edward Venables Vernon Esquire, also had interests in some of the land. This portion of the lands and manor of Clontarf which had been demised to the trustees of the late Thomas Popham Luscombe involved two separate plots. The first amounted to 5a. 0r. 23p. (Irish measure) to be held for nine years from 29 September 1860 at an annual rent of £29 9s. 6d. The second was held on a ninety-one year lease from 29 September 1778 and amounted to 16a. 0r. 4p., excepting the portion which had subsequently been taken by the Dublin and Drogheda Railway Company. In addition to an annual rent of £92 6s. 2d., there was a covenant by the lessees to provide a labourer and a cart and harness for six days every year (or 13s. a year in lieu thereof). There was a further stipulation not to erect or build any cottages on the said demised premises, without consent in writing.

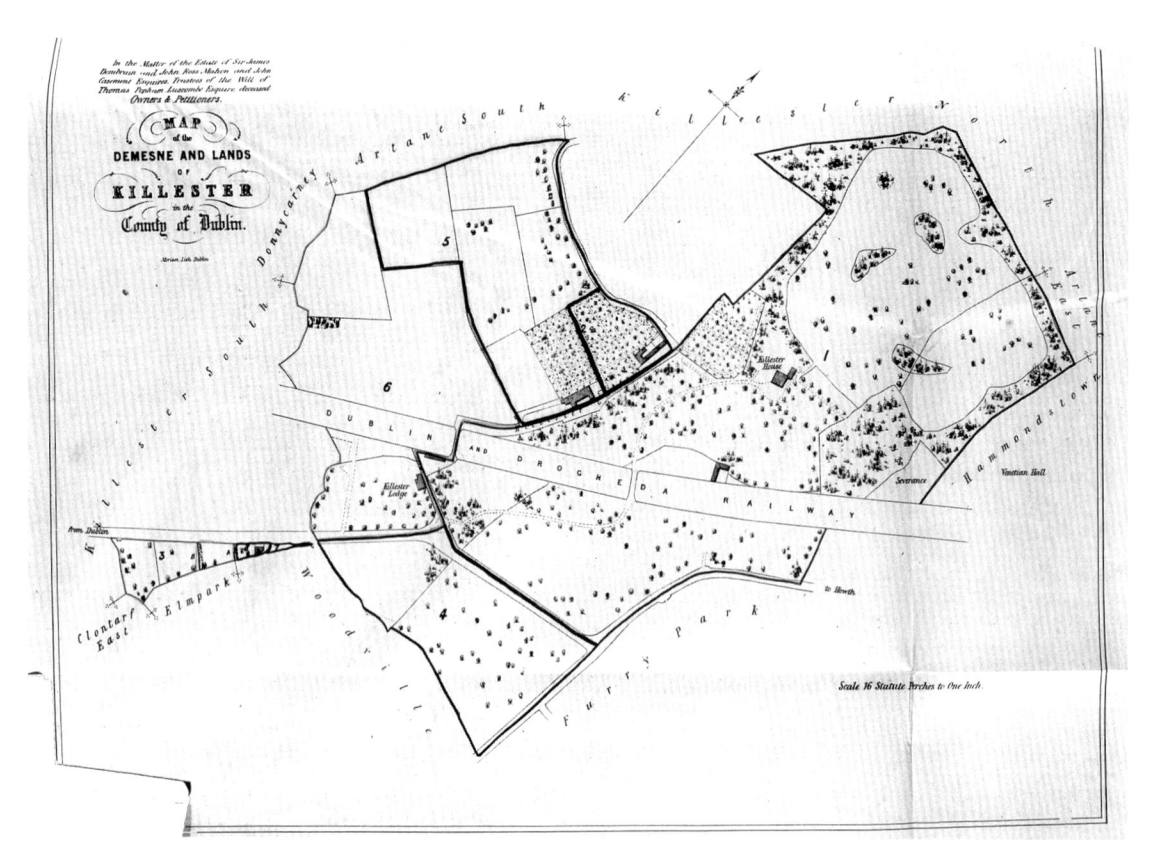

A simplified version of the table which corresponds to the map reproduced above accounts for the mansion house and demesne lands of Killester, and other portions of said lands, held under several leases for terms for years, and 1a. or 18p. statute measure, part of the lands of Furrypark, now adjoining part of said demesne lands, held in fee, amounting to 117 acres 1 rood and 37 perches statute measure:

1 Mansion house and demesne lands	William Lynch MD	£250 rent	64a. 1r. 28¼p.
2 House and garden [Killester Gardens]	Patrick Kirby	£25 rent	2a. 1r. 37p.
3 Lands known as Silver Field	Mrs Fortune (rep. Thos Hope)	£11 1s. 7d.	2a. 2r. 7½ p.
4 Other part of said lands	Joseph Bradley	£41	12a. 0r. 27p. (tenant from year to year)
5 Other part of said lands	Thos Alexander MD (rep. Robert Mayne)	£64 12s. 4d.	16a. 3r. 25p.
6 Other part of the lands	Dublin & Drogheda Railway Co.	£138 1s. 4d.	18a. 2r. 18p.
7 Other parts adjoining Silver field	John Bingham	peppercorn	0a. 0r. 26p.
8 Other parts adjoining Silver field	John Pennefather	peppercorn	0a. 0.r 28½p.

17 Parish map, 1863. Extract from Ordnance Survey plan, 1:2,500, Co. Dublin sheet XIX(1), 1863 edition. [FSLA]

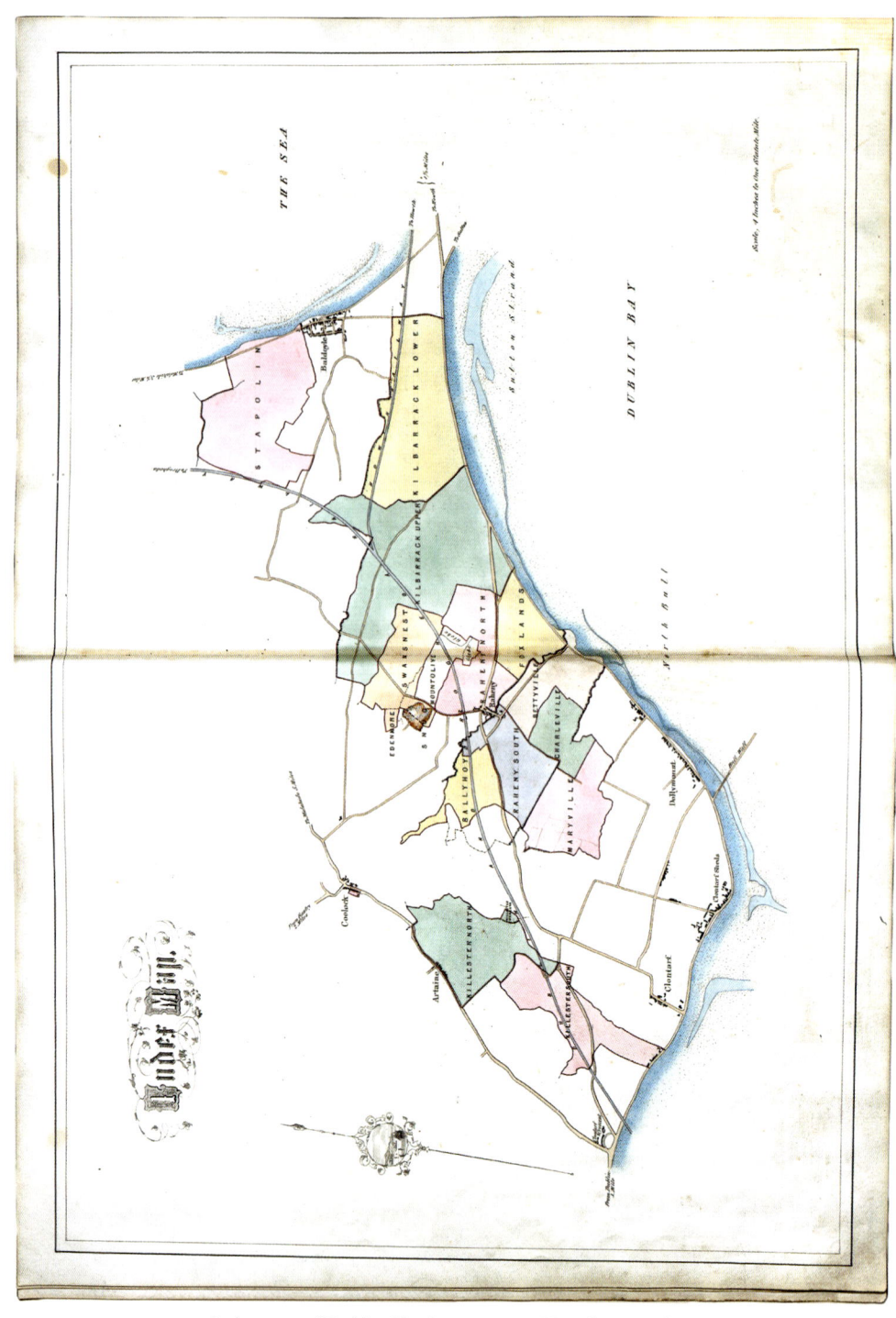

18 Index map of Dublin North properties. Howth estate. [FSLA]

More information is available for Killester from the mid-nineteenth century than from previous periods. In addition to the landed estates rental and the 1863 parish map discussed above, the Howth estate undertook a survey of their properties, which resulted in the production of a series of beautiful maps which are published here for the first time.

The index map to the Howth estate demonstrates the extent of holdings in 1863. Individual townlands are named, including Killester North, Killester South and Killester Demesne. Villages and clusters of housing are also indicated, as at Artane, Baldoyle, Clontarf, Coolock, Dollymount and Raheny. The surveyor has not chosen to indicate any such cluster at Killester, suggesting that there was limited concentration of housing in the area at that time.

The Howth estate held the entire townland amounting to 95 acres 1 rood and 5 perches. The substantial residences are detailed: Oatley (later renamed Craigford), Mount Dillon at the northernmost end of the townland, Killester Abbey, and Killester House; the gardens, wooded areas and parkland associated with the demesne house are shown, as is the nun's walk, the ice house and various lodges and farm buildings. The church (in ruins) and graveyard are marked (number 32 on the map), while there are buildings at Killester Gardens which was technically located in Killester South but adjoining the concentration of activity associated with Killester House. The crossing points over the Dublin and Drogheda railway are shown. The linear 'village' at Artane can also be seen. It should be noted, of course, that any property on the west side of the Malahide Road and therefore outside of the ownership of the Howth estate is not depicted. Individual plots are detailed in the index, which lists the name of the tenant and the size of the holding in both Irish and English acres. Just three tenants were listed: Thomas Alley was in possession of more than half the townland (numbered 1 to 19 on the map) including the cottages in Artane and the two substantial residences named Mount Dillon and Oatley. These would have been sub-let, but the names and further details of their occupants are not available. The late Sir William Gleadowe Newcomen's representatives were still listed in possession of just over 23 acres, including Killester House and its associated grounds (numbers 24 to 31 on the map), while Daniel Nugent held a smaller holding of slightly more than 9 acres, as representative of a Miss O'Callaghan (numbers 20 to 23 on the map). This centred on Killester Abbey.

The estate map for Killester South provides a more detailed depiction of Killester Gardens and Killester House (the second, not the demesne house) and grounds. Both dwellings were part of a single unit amounting to 2 acres 3 roods and 4 perches, number 1 on the map. Extensive planting and a number of out-buildings are evident. The first ten numbered plots in Killester South, which also included Killester Lodge and Silverfield House, were in possession of the representatives of Sir William Gleadowe Newcomen, amounting to 22 acres 1 rood and 18 perches in total. Also

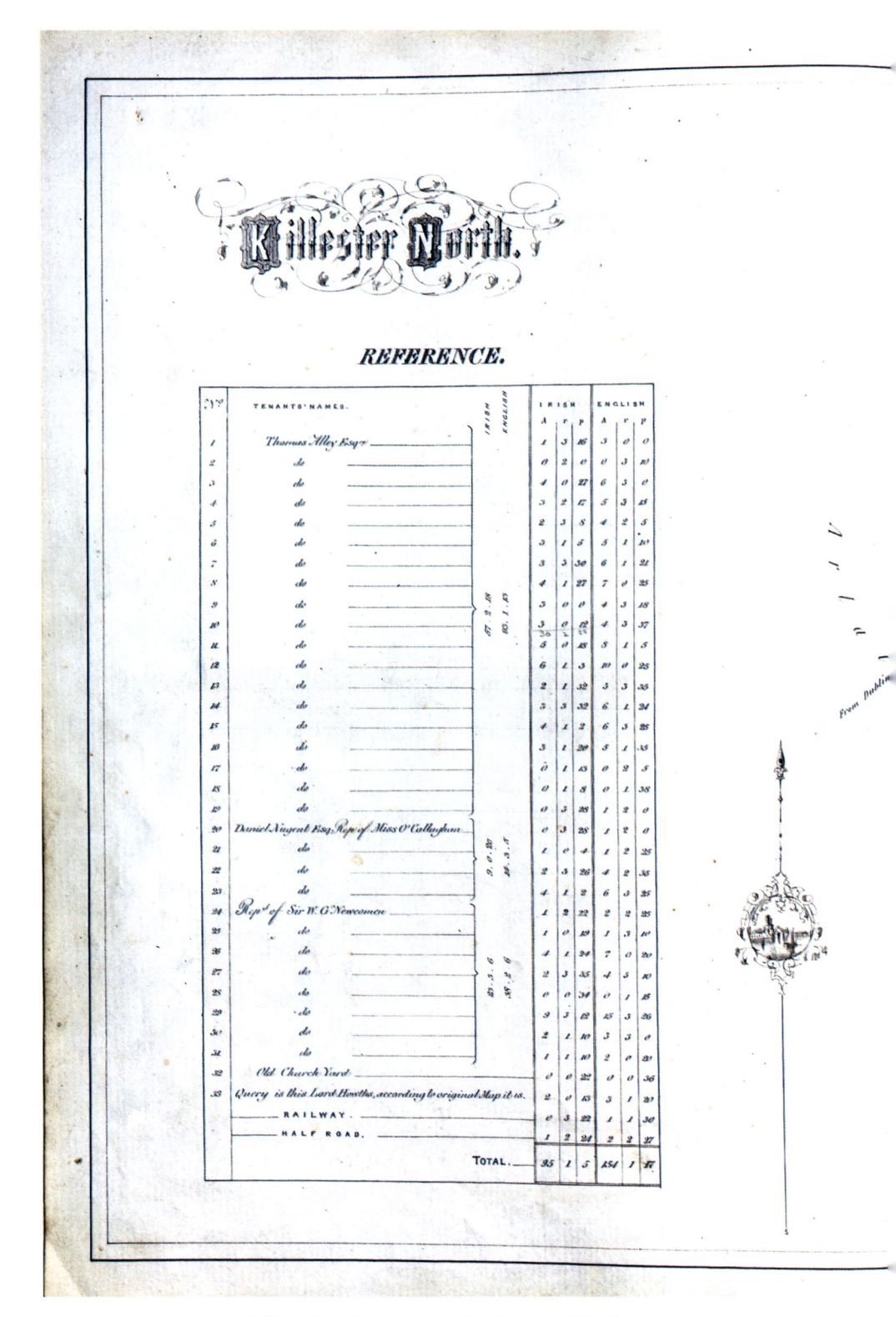

Killester North.

REFERENCE.

Nº	TENANTS' NAMES.	IRISH	ENGLISH
		A r p	A r p
1	Thomas Alley Esqr	1 3 16	3 0 0
2	do	0 2 0	0 3 10
3	do	4 0 27	6 3 0
4	do	3 2 17	5 3 15
5	do	2 3 8	4 2 5
6	do	3 1 5	5 1 10
7	do	3 3 30	6 1 21
8	do	4 1 27	7 0 25
9	do	3 0 0	4 3 18
10	do	3 0 12	4 3 37
11	do	5 0 18	8 1 5
12	do	6 1 3	10 0 25
13	do	2 1 32	3 3 35
14	do	3 3 32	6 1 24
15	do	4 1 2	6 3 25
16	do	3 1 20	5 1 35
17	do	0 1 13	0 2 5
18	do	0 1 8	0 1 38
19	do	0 3 28	1 2 0
20	Daniel Nugent Esq. Reps of Miss O'Callaghan	0 3 28	1 2 0
21	do	1 0 4	1 2 25
22	do	2 3 26	4 2 35
23	do	4 1 2	6 2 5
24	Reps of Sir W. G. Newcomen	1 2 22	2 2 25
25	do	1 0 19	1 3 10
26	do	4 1 24	7 0 20
27	do	2 3 35	4 3 10
28	do	0 0 34	0 1 15
29	do	9 3 12	15 3 26
30	do	2 1 10	3 3 0
31	do	1 1 10	2 0 25
32	Old Church Yard	0 0 22	0 1 36
33	Query is this Land Howths, according to original Map it is.	2 0 13	3 1 30
	RAILWAY	0 3 22	1 1 30
	HALF ROAD	1 2 24	2 2 27
	TOTAL	95 1 5	154 1 17

19 Killester North estate map. Howth estate. [FSLA]

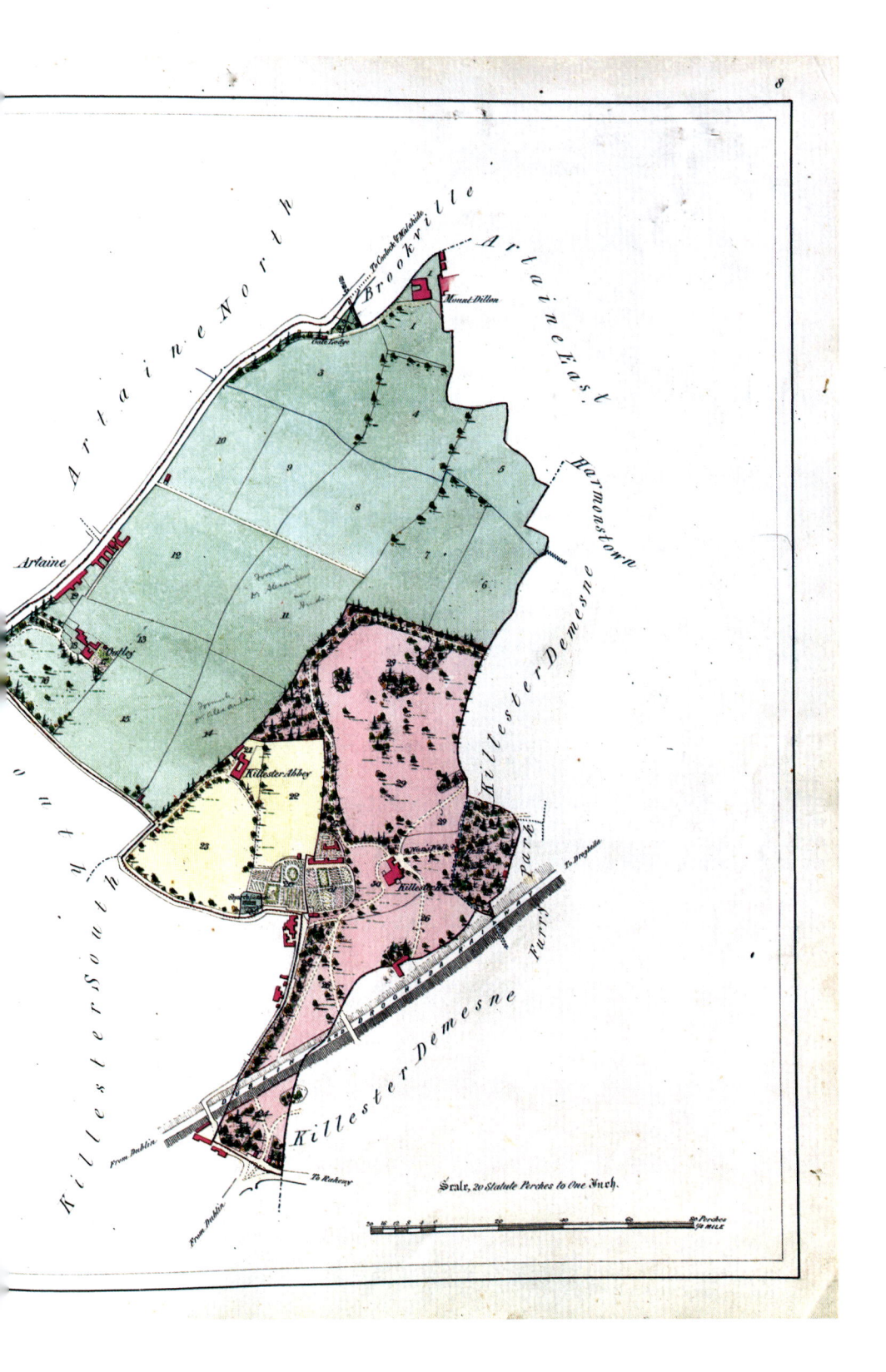

Artaine North
Brookville
Artaine East
Mount Dillon
Harmonstown
Artaine
Killester Demesne
Killester Abbey
Killester Park
Killester South
Killester Demesne
From Dublin
To Raheny
From Dublin
To Drogheda

Scale, 20 Statute Perches to One Inch.

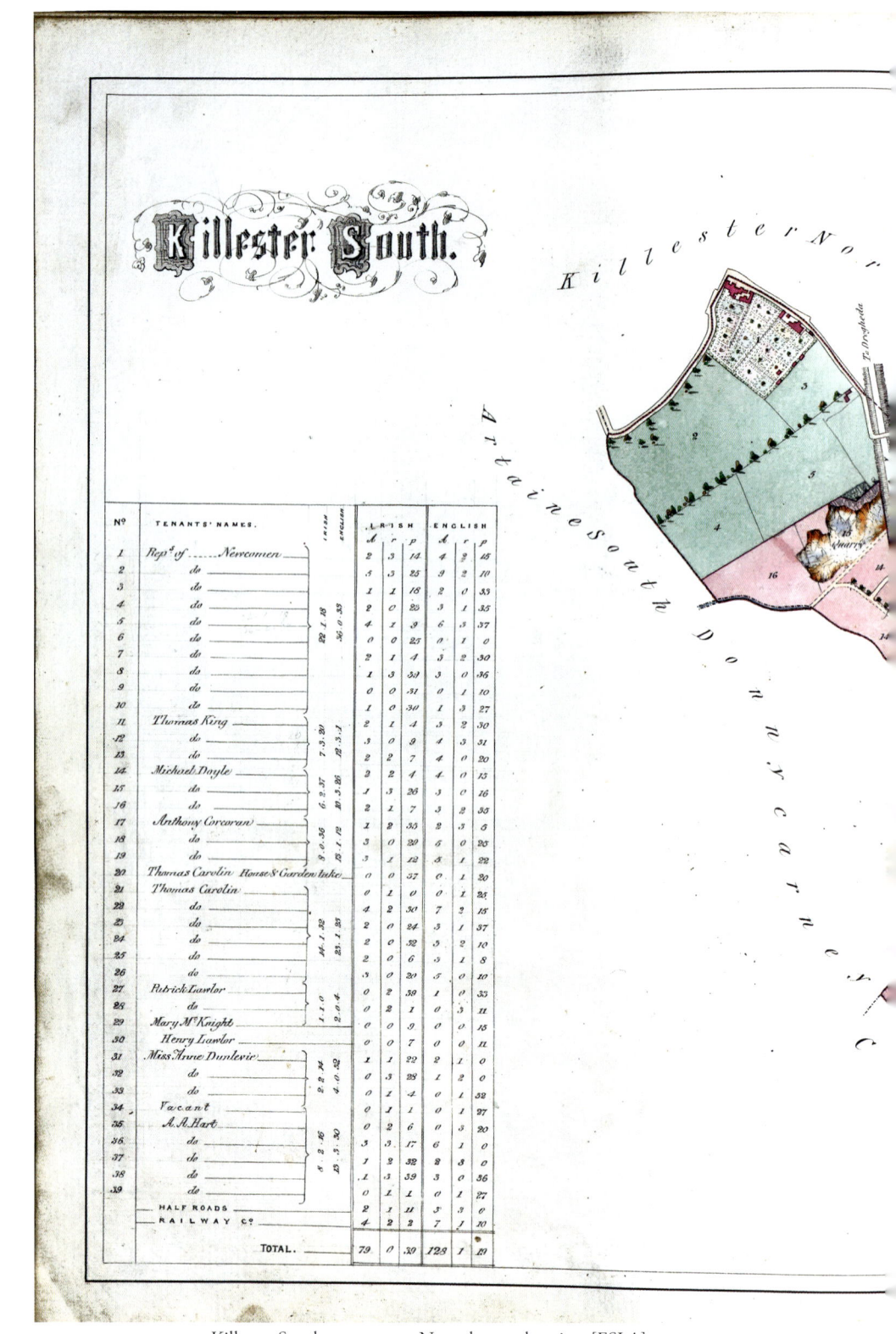

Killester South.

Nº	TENANTS' NAMES.		IRISH			ENGLISH		
			A	R	P	A	R	P
1	Repᵈ of Newcomen		2	3	14	4	3	15
2	do		5	3	25	9	2	10
3	do		1	1	16	2	0	33
4	do	22.1.18	2	0	20	3	1	35
5	do	36.0.33	4	1	19	6	3	37
6	do		0	0	25	0	1	0
7	do		2	1	4	3	2	30
8	do		1	3	39	3	0	36
9	do		0	0	31	0	1	10
10	do		1	0	30	1	3	27
11	Thomas King		2	1	4	3	2	30
12	do	7.3.20	3	0	9	4	3	31
13	do	12.3.1	2	2	7	4	0	20
14	Michael Doyle		2	2	4	4	0	13
15	do	6.2.37	1	3	26	3	0	16
16	do	10.3.25	2	1	7	3	2	35
17	Anthony Corcoran		1	2	35	2	3	6
18	do	9.0.36	3	0	20	6	0	25
19	do	15.1.12	3	1	12	5	1	22
20	Thomas Carolin House & Garden		0	0	37	0	1	20
21	Thomas Carolin		0	1	0	0	1	25
22	do		4	2	30	7	2	15
23	do	14.1.32	2	0	24	3	1	37
24	do	23.1.33	2	0	32	3	2	10
25	do		2	0	6	3	1	8
26	do		3	0	20	5	0	10
27	Patrick Lawlor		0	2	39	1	0	33
28	do	1.1.0	0	0	2	0	0	11
29	Mary McKnight	2.0.4	0	0	9	0	0	15
30	Henry Lawlor		0	0	7	0	0	11
31	Miss Anne Dunlevie		1	1	22	2	1	0
32	do	2.2.24	0	3	28	1	2	0
33	do	4.0.12	0	1	4	0	1	32
34	Vacant		0	1	1	0	1	27
35	A. A. Hart		0	2	6	0	3	20
36	do	8.2.16	3	3	17	6	1	0
37	do	13.3.30	1	2	32	2	3	0
38	do		1	3	39	3	0	36
39	do		0	1	1	0	1	27
	HALF ROADS		2	1	11	3	3	6
	RAILWAY Cº		4	2	2	7	1	10
	TOTAL.		79	0	39	129	1	21

20 Killester South estate map. Note the north point. [FSLA]

Killester Demesne

Elm Park

Clontarf East

Clontarf West

From Dublin

To Dollymount & Howth

DUBLIN BAY

Scale, 20 Statute Perches to One Inch.

shown on this map is the quarry just west of the railway line (number 15). Overall, there were more individual tenants in Killester South than in Killester North. The influence of the Howth Road which ran through this townland is more in evidence. See, for example, at number 18 where the original plot has been subdivided and three substantial dwellings have been built, each with a sweeping driveway up to the entrance.

LAST RESIDENTS OF KILLESTER DEMESNE

The brief mention of Killester, within the entry for Raheny in Thom's directory of 1868, described it as a parish of 279 acres, with a population 456, 'through which the railway passes at a depth of excavation of 36 feet and where it is crossed by an iron suspension bridge, 84 feet in span' (Thom's, 1868). It seems that the railway cutting was the most remarkable thing to note about the area by that time. Killester retained its rural, unspoiled character, despite the intrusion of the railway – indeed, the railway cuttings proved to be a useful habitat, where varieties of orchids could be found growing in profusion in the 1890s (Colgan, 1895: 194, 197). The 1868 Thom's listing yields a dozen names in total, relating to ten named residences. The most significant, with the highest rateable valuation at £135, was Killester Demesne, occupied by Lieutenant A.H. Lynch, while a separate dwelling, 'Killester Hall', was occupied by Arthur F. Lynch of the Army Medical Staff: the Lynch family had been in occupation for ten years at this point. Under a lease dated 24 April 1858 made by Thomas Charles Popham Luscombe esq. (who died in June of the same year) to William Lynch MD, the mansion house and lands of Killester, including a one-acre portion of the townland of Furry Park held in fee, known as the demesne lands of Killester, amounting to 64 acres 1 rood and 28¼ perches, were held at a yearly rent of £250.

The Roman Catholic Lynch family is the last recorded to have lived long-term in Killester Demesne. William J. Lynch's (FRCSI) large family was in occupation by 1858. His eldest daughter married from there in 1863 (*Evening Freeman*, 13 November 1863) while his fifth son, Dr Robert Lynch, physician and surgeon, died there in 1884 at the age of forty-four. Lily, daughter of the late Charles Edgeworth Lynch, was married in 1892. Her father Charles, sixth son of William, had died in 1882 after a long illness. Charles had been a member of the Papal Zouaves, an infantry battalion of mainly young, unmarried and Catholic volunteers who assisted Pope Pius IX in his struggle against the Italian unificationist Risorgimento in the 1860s. Another family member living at Killester Demesne, Dr Arthur Henry Francis Lynch, held Crimean, Turkish and New Zealand war medals and clasps 'for services in the trenches and field' according to the Irish medical directory of 1875.

An interesting court case concerning Dr William Lynch and the premises at Killester was reported in the *Freeman's Journal* in 1867, which casts some doubts on

The evolution of Killester from earliest times

the use to which the house was being put at that time. The report suggests that Dr Lynch, who ran Heartfield (aka Hartfield) Private Lunatic Asylum in Drumcondra, had leased Killester House for the use of a patient, Mr Thomas Lahiff, of Galway, who had long-term dementia. Lynch had come under scrutiny having taken out a very large insurance policy (£1,000) on the life of his patient. During the proceedings it was admitted that Dr Lynch had taken out the policy, but this was justified on the grounds that 'that was done to secure a very heavy expense which Dr Lynch incurred in renting and supporting the establishment at Killester for the lunatic's use' (*Freeman's Journal*, 22 July 1867). Whatever the circumstances in the 1860s, it would appear that the Lynch family occupied the main house (Killester Demesne) and Killester Hall from the 1860s through to the late 1880s. Dr William James Lynch himself died at Killester Demesne on 10 November 1887, aged 86.[8] By the time that the 1889 edition of Thom's directory was published, Killester Demesne was listed as being vacant. Garrett (2006: 13) suggests that after the Lynch occupancy, Lord Howth employed a caretaker to look after the property.

An *Irish Independent* columnist in 1903 remarked that there were 'some interesting ruins of a Cistercian abbey in the grounds of Killester', although there does not appear to be any evidence to support the assertion that there was a Cistercian association with the area. 'The last tenant of this place was Dr Lynch, well and for many years known around this neighbourhood as a mental specialist. Since his death, this beautiful house and grounds are empty, a prey to waste, not having been taken for some charitable institution—the fate which seems to be swiftly overtaking all such dwellings near Dublin' (*Irish Independent*, 28 September 1903).

KILLESTER AT THE TURN OF THE TWENTIETH CENTURY

The 1901 census provides a snapshot of life in the Killester area. The twenty-three dwellings of Killester North were enumerated within Drumcondra Rural for census purposes, while eleven houses in Killester South were recorded as part of Clontarf West. A separate entity described as Killester Demesne was located within the Clontarf district for census enumeration purposes. Only two dwellings containing nine people, from two families, were living within the boundary as recorded in the census. The houses were leased from the Rt Hon. the Lord of Howth and contrasted significantly. The larger of the two was a first-class house with fourteen rooms, occupied by the Thompson family. By cross-reference with Thom's directory, this appears to be the house named Woodville which was present on the first edition Ordnance Survey map and survived until at least the 1930s. Woodville was located just north of present-day Castle Grove. Associated with the house there were a number of out-houses, including a stable, coach-house, cow-house, piggery, fowl-house and shed. Widower William Thompson, a retired farmer who had been born in Co. Down,[9] lived there with his

Dublin-born children, widowed daughter Maria Magee and his son, also William Thompson, a farmer. The family belonged to the Church of Ireland, whereas the larger Mills family were Roman Catholics. Bridget Mills,[10] the head of the family, was a 55-year-old widow whose occupation was described as a 'housekeeper', suggesting that she worked in one of the larger nearby residences. Her daughter Mary was a laundress, while four unmarried sons ranging in age from twenty-three to thirty-five years were all builder's labourers. All six adults were crowded into a third-class two-roomed dwelling. The majority of the housing in Killester North fitted the profile of the Mills family residence, being small in size but sometimes containing large families. Of the twenty-three dwellings listed, six were two-roomed cottages, three had three rooms, nine had four rooms and there was just one five-roomed house. The majority of these cottages were along the Malahide Road close to the Kilmore Road junction, where the combined post office and shop was also located. Three larger houses were listed in 1901, with ten, twelve and nineteen rooms respectively, while three dwellings were uninhabited at the time of the census.

Ten years later, when the 1911 census was undertaken, the demesne was enumerated within the Drumcondra Rural district. Again, there were only two occupied houses recorded for Killester Demesne, but this time both were small two-roomed cottages, and these were leased from Margaret Dunne of Harmonstown. General labourer Terence Taaffe,[11] his wife and niece lived in one, while Bernard Byrne,[12] also a general labourer, lived in the other with his wife Kate and five children. Each of the cottages also had a piggery and fowl-house, suggesting one way in which the families supported themselves. The profile of residences in Killester North had changed little since the previous census in 1901, although the overall number of houses enumerated had increased from twenty-three to thirty-two. Of these, seven were two-roomed cottages and twenty had three rooms. Typical of the inhabitants were James Farrell, Joseph Teeling and James Foody, all of whom were agricultural labourers. The houses on Killester Lane and at Quarry Cottages were listed within the Clontarf West ward. There were three substantial first-class dwellings on Killester Lane, the Kirby family's Killester Gardens, retired accountant Thomas Hall's Killester Park, and engineer Frederick E. Cairns and his young family who resided at a newer dwelling known as Killester House with four servants. Despite what the name might suggest, the four Quarry Cottages were all occupied by railway workers, including a ganger and three railway plate layers.

As the mansion house at Killester was unoccupied at the times of the census in 1901 and 1911, no information was recorded in relation to it. As late as 1911, the Howth estate office was advertising the property to be let on a long lease or sold (*Evening Irish Times*, 20 November 1911). The possibility of buying out the fee simple of the demesne was under consideration in 1919, suggesting that it was still owned by the Howth estate at this date (*Dublin Evening Telegraph*, 22 March 1919) but there is

The evolution of Killester from earliest times

21 Venetian window joinery salvaged from Killester House and used to frame a fireplace in the entrance hall at Howth Castle, *c.*1909. Photograph: Rob Goodbody.

22 Chimneypiece salvaged from Killester House and reused in Lutyen's new library at Howth Castle, *c.*1909. [DCC]

an alternative view, discussed in the chapter that follows. Meanwhile the Howth estate would appear to have had more success in relation to the neighbouring Killester Abbey, advertising the house, large garden and about thirteen acres of 'excellent ground' (as Killester Abbey, Artane) either on a 999-year lease or for sale in fee simple in 1913 (*Evening Irish Times*, 2 August 1913).

Clearly the Killester mansion house had fallen into dereliction by the time that the last earl of Howth, William Ulick Tristram St Lawrence, died in 1909. On his death, the family titles became extinct as he had no male heir, and the property was inherited by a nephew, Julian Gaisford. The latter sold his house in England and commissioned the celebrated English architect Edwin Lutyens to update and add to Howth Castle. As part of this project, a significant amount of material was salvaged from the house at Killester and reused in Howth. This includes the very fine early eighteenth-century stone chimneypiece from Killester House, which was removed and reused in the library at Howth Castle, housed within the Gaisford Tower, a substantial new addition by Lutyens at the end of the west wing. Other features from Killester that were salvaged include Venetian window joinery, which was used to frame a fireplace in the entrance hall, mahogany doors and several bedroom fireplaces. When the contents of Howth Castle were sold in 2018, a number of elements from Killester House were noted in the catalogue. The salvage of materials from the house in 1910–11 saved these fragments of its former grandeur from ongoing dereliction and the fire which would engulf it less than a decade later.

FINAL YEARS OF KILLESTER DEMESNE

The history of the Killester Demesne was not untypical until its final years. A look at large-scale maps of Dublin for the beginning of the twentieth century shows that there were many such houses with demesnes dotted around the landscape, beyond the city boundaries. Their circumstances and condition varied greatly, in line with the circumstances of their owners. Marino House, no more than two kilometres from Killester, was demolished during the building of Dublin Corporation's model housing scheme in the 1920s. Although the earlier plans had provided for its preservation, it seems to have deteriorated to such an extent by the 1920s that demolition was inevitable. Many other houses also disappeared as the city encroached on them. This is a process which continued throughout the twentieth century. University College Dublin's campus comprises the demesnes of a number of houses in the Donnybrook area which had fallen into various states of decline by the 1930s and were easy to obtain. It was, however, not usual for Dublin Corporation to take possession of a house and its demesne. The circumstances in which this happened in Killester are explored in detail in the following chapter. It resulted in the building in the early

1920s of one of the most distinctive housing schemes for ex-servicemen, often referred to as a garden village or garden suburb. This might also have been instrumental in the destruction of the house by fire in May 1920.

The burning of houses during the revolutionary period 1919–23 was generally associated with the destruction of the country houses of the aristocracy and landed gentry, but not limited to this. The old mansion known as Killester House fell victim to arsonists on the morning of 20 May 1920. On Wednesday 26 May 1920, the *Pall Mall Gazette* ran a front-page headline 'Ireland's newest fire orgy'. The accompanying article detailed premises that had been burned in what it termed the continuing 'Irish campaign of fire', including Kilbrittain Castle, Co. Cork, a farm and stockyard near Ballinasloe, Waterville courthouse in Co. Kerry, Glendallough (*sic*) House (the property of the Department of Agriculture) and Killester House (the property of Dublin Corporation), involving £2,000 damage. On the same date, under the headline 'a catalogue of outrage', the *Irish Times* listed events across the country, ranging from destruction of letter-boxes and cattle drives to thefts, armed raids and attacks upon individuals, shootings and burnings. A paragraph within this long list (p. 7) describes the complete destruction by fire of Killester House, noting that the house was unoccupied, and asserting that 'it was destroyed to prevent the military from taking it over'. The limited newspaper coverage which the destruction of the house at Killester garnered can be understood within the context of the Civil War period when it was but one of many 'outrages'.

It is not clear exactly why Killester House was targeted, but this was the end of an era. The rubble remained at the scene for some time. In July 1920, an eleven-year-old local boy, Brendan O'Brien, died in a fall there, having followed a goat into the ruins (*Freeman's Journal*, 19 July 1920; General Register Office). This may have prompted the removal of the remaining walls of the house. A raised area in the grassy parkland area at Middle Third, north of the former Legion Hall, may be due to the clearing of the site following the fire. The house was never rebuilt. All that would survive would be the unusual district name 'Demesne'. Most present-day residents of the area are completely unaware of the long and eventful existence of the manor house at Killester.

CHAPTER TWO

The Killester Garden Village

JOSEPH BRADY

HOMES FIT FOR HEROES

When the First World War began there was an urgent need to swell the numbers of the British army and an intensive recruitment campaign began in Ireland which quickly bore fruit. The *Irish Times* reported in October 1914 that over 5,800 recruits had been signed-up in Dublin alone since the war began (26 October 1914: 6). Doubtless their motivations were varied. Some joined out of a sense of duty to the empire, others because they were exhorted to do by John Redmond and who believed that it would copper-fasten home rule. Others joined because it was regular income, if not particularly high, and there was a chance of learning a trade. In the absence of conscription in Great Britain all forms of persuasion were used. 'Your King and Country want you' was a popular song from 1914 and was sung with enthusiasm by Helen Clark (visit Youtube for a recording). The Order of the White Feather sent many to the trenches. This involved women handing white feathers to men of military age who were not in uniform. In fact, so effective was the symbolism that civil servants and others engaged in work of national importance had to be given badges proclaiming that they were working for king and country to avoid harassment. A wide range of posters appealed to all of the motivations – friendship, the glory of Ireland, the treatment received by 'Catholic' Belgium, courage, duty and home rule, as captured by this poster (Figure 23).

Henry Lefroy, later Major, from Killaloe was tasked with recruitment for the Royal Munster Fusiliers and the Royal Irish Regiment. During his recruitment activities, he often heard the complaint that men were being asked to go to fight for land in another country when they owned nothing at home. At the end of 1915, he was appointed personal assistant and staff officer to the Inspector General R.A. and this gave him access to the upper echelons of the British army. This began his lifelong campaign for land and accommodation for ex-servicemen. His was not a lone voice. There had long been efforts in Great Britain to settle ex-servicemen on the land in the hope of increasing land productivity and rural population and reducing national unemployment and urban congestion (see Aalen, 1988). The war put new energy into this movement. When it ended there would be hundreds of thousands of men seeking employment and a rural settlement programme seemed an excellent way of improving food security, which had been exposed during the war, as well ensuring social stability.

The Killester Garden Village   47

23 First World War recruitment poster. John Redmond exhorting men to join up. [PC]

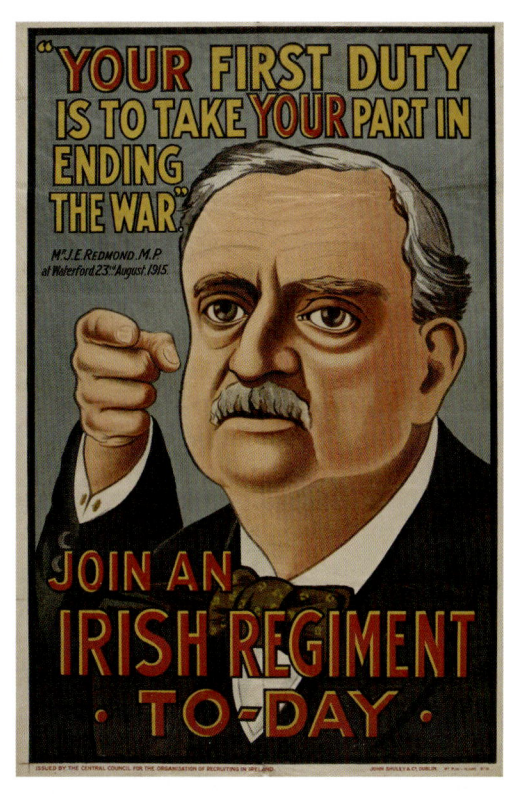

The focus was initially on rural settlement but it gradually became meshed with a wider programme to improve the housing of the working classes. Recruiters were said to have been shocked at the physical condition of many of the recruits from the urban slums.

Work was under way preparing suitable legislation to facilitate better working-class housing as the war was coming to an end. In July 1917, Sir John Tudor Walters was asked to head a committee 'to consider questions of building construction in connection with the provision of dwellings for the working classes in England and Wales, and report upon methods of securing economy and despatch in the provision of such dwellings'. Its remit was later extended to include Scotland and its report was published in November 1918. The rather anodyne terms of reference gave no indication of how significant a report this was. Unusually for government reports, it was a detailed and practical analysis of how to build suburbs and it would set the parameters for social housing in the United Kingdom and elsewhere for the next twenty years. Its principles were heavily influenced by the garden city movement, at least to the point that housing would be of good quality at low density and with an emphasis on green space.

A specific focus on ex-servicemen was strengthened when, having called a general election on 12 November 1918, David Lloyd George made a promise during an election speech in Wolverhampton on 24 November 1918 to make 'Britain a fit country for heroes to live in' and went on to say 'slums are not fit homes for the men who have won this war or for their children' (MacArthur, 1992: 69–70). However, these sentiments were collapsed into the much more catchy phrase – 'homes fit for heroes' – now remembered as the actual quotation.

This did not resonate immediately in Ireland where the debate was on home rule or independence but a similar promise was already in the air. A pressing need for more troops in France prompted the UK prime minister Lloyd George to introduce legislation for conscription in Ireland in 1918. A suggestion that this would be linked to the introduction of home rule alienated both unionists and nationalists but perhaps the most significant opposition came from the Catholic hierarchy who declared that it was an oppressive and unjust law and, even more dramatically, called on all its followers to oppose it with every means which were consistent with the laws of God (see Foster, 1992: 202). Fortunately for the authorities, a change in the fortunes of the war and the accelerated arrival of troops from the United States allowed them to drop the idea quietly. However, they still persisted with recruitment and Lord French, the new lord lieutenant, issued a proclamation, dated 3 June 1918, which sought the voluntary recruitment of 50,000 men. Clause 5 of the document made the offer: 'We recognize that men who come forward and fight for their motherland are entitled to share in all that their motherland can offer. Steps are, therefore, being taken to ensure, as far as possible, that land shall be available for men who have fought for their country, and the necessary legislative measure is now under consideration'.

THE LEGISLATION

Ireland would require its own legislation to give effect to any promise for ex-servicemen but the UK government was true to its word. The first attempt at legislation in 1918 did not find favour and was withdrawn. It was therefore late in 1919 before the House of Commons debated and ultimately approved the Irish Land (Provision for Sailors and Soldiers) Act, 1919. As explained in the House of Commons by the chief secretary for Ireland, Mr Ian MacPherson, on 18 November, the legislation was 'a generous attempt to give to those Irishmen who fought for their country facilities to settle on the land'. There would be two kinds of applicants. The first were those who wished to be full-time farmers while the second were men 'not desirous of devoting his whole time to agriculture but anxious to employ himself in another industry, while at the same time wishing to have an allotment with security of tenure and a house in which to live'.

The allocation of farms would be handled by the Land Commission under the land purchase acts, drawing on the land available to the estates commissioners and the Congested Districts Board. Ex-servicemen would be given priority access and would

The Killester Garden Village

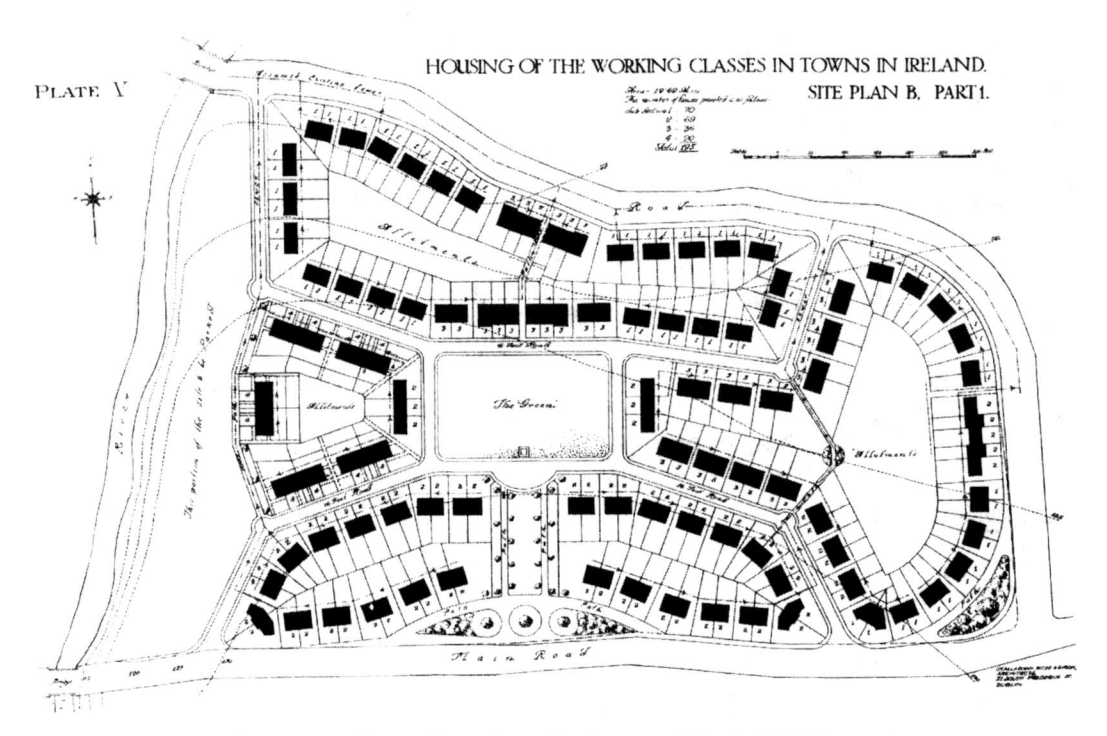

24 One suggested layout for working class houses in Ireland. [HMSO, 1919]

ultimately become owners. The other category of ex-servicemen would be looked after by the Local Government Board using powers under the Labourers' (Ireland) Act of 1906. The latter permitted local authorities to build houses for agricultural labourers and to give them a plot of ground of up to one acre. In this case, the task was given to central authority, the Local Government Board for Ireland, and the size of the plot was increased to a maximum of two acres. It was envisaged that this would be a time-limited process and the Board's power to acquire property for this kind of housing would lapse after three years. Those getting land would be tenants, there was no suggestion at this point that ownership would follow.

It was intended that supports would be provided to both full-time and part-time farmers via the Department of Agriculture and it was the hope of the chief secretary of 'seeing large colonies of these soldiers scattered all over Ireland, and I think it will be of very great value to them to have the Board of Agriculture and Technical Instruction at their disposal'. The use of the term 'colony' raised some hackles in Republican circles, and it was used both positively and negatively over the years. The chief secretary saw another value in grouping these men into colonies. As he put it: 'I also look at the establishment of these colonies from the social point of view. Whatever our feelings may be about Ireland, it is true that these men have had since their return in many parts of the country a very difficult time. Their association together in a colony of this kind will not only be of material value to them, but will afford them a great amount of coherent sympathy and protection.'

The legislation got an easy passage through the UK parliament and received royal assent on 23 December 1919. Building began immediately. In parallel with these plans, the LGB had sponsored a competition for the design of houses for the working classes generally (not just ex-servicemen) and published their recommendations, together with examples of designs and layouts, in 1919. Though these were not templates, Killester's layout is very much in harmony with these ideas.

KILLESTER GARDEN SUBURB/CITY

Ebenezer Howard's ideas on garden cities as an alternative to the increasing sprawl of Victorian cities were still relatively novel, having been first published in 1898. Instead of more and more urban sprawl, he proposed that society be organized into a connected system of small urban centres, with limited populations, which had a close connection to the surrounding rural landscape. He was one of many urban reformers in the nineteenth century and his ideas might have had limited influence except that he was part of an influential set that included publishers and politicians. His ideas were given a wider audience and developed by people destined to become world-renowned experts on town planning. Two people, in particular, became very involved in Dublin's urban evolution: Raymond Unwin and Patrick Geddes. Unwin and his partner Barry Parker had designed the first garden city at Letchworth in 1904 but their development of the concept of the 'garden suburb' was perhaps more significant. Detractors took the view that this was a very significant watering-down of Howard's ideas but garden suburbs were easier to achieve than garden cities. Unwin and Parker designed what is regarded as the first garden suburb in Hampstead (London) in 1906. It was low-density, tree lined with roads respecting the landscape and much open space and woodland. Unwin went on to be an influential member of the Tudor Walters committee and though the report's provisions did not directly apply to Ireland, its endorsement of garden city principles was quickly recognized in Dublin both by Dublin Corporation and the Local Government Board.

Patrick Geddes was a pioneering Scottish sociologist, geographer, philosopher and town planner who advocated an holistic approach to urban planning, emphasizing the importance of considering social, environmental, and cultural factors in the design of cities and the production of a plan. His approach was evidence based, advocating a detailed survey before planning and arguing that any plans developed needed to be flexible and adaptable.

One of his projects had been the development of an exhibition entitled 'Cities and Town Planning'. With Edinburgh as a central element it demonstrated the evolution

The Killester Garden Village

25 Present-day aerial view of housing in Hampstead Garden Suburb. [Google Earth]

and development of towns and cities globally with the aim of showing the possibilities of civic planning. In 1911 he was invited to bring the exhibition to Dublin by the lord lieutenant, Lord Aberdeen. In reality, while Lord Aberdeen was an enthusiastic supporter of civic reform, his wife, Ishbel, was even more energetic in her championing of charitable causes, social reform and home-grown Irish industrial development. The exhibition's assistant director was a young architect planner called Frank Mears, who had been centrally involved in the development of the exhibition and he was sent to Dublin to make the practical arrangements. It was during this and later stays in Dublin that he got to know the influential and energetic people involved in civic improvement, especially E.A. Aston.

Aston was, for many decades, an enthusiastic promotor of civic transformation in all its aspects. His obituary in the *Irish Times* in 1949 described him as a 'pioneer and crusader' (7 March 1949: 5). He had a deep interest in housing reform and town planning and was a member of many campaigning organizations. He was also a developer who attempted to build housing that reflected his ideals (see below). He was one of the founders of the Housing and Town Planning Association of Ireland and that alone would have brought him into contact with Mears.

The Civic Exhibition opened its doors in the Linenhall near King's Inns Street on 15 July 1914, enhanced with demonstrations and a series of public lectures. An international competition for a town plan was held in association with the exhibition. This offered a very substantial prize of £500, donated by the Aberdeens, for the 'best design of a plan for the improvement and extension of Dublin'. The adjudicators were John Nolen of Massachusetts (US), Patrick Geddes and Charles McCarthy, the city architect. Nolen had visited Dublin to advocate in favour of town planning. Geddes was a logical choice as one of the adjudicators given his role in the exhibition, his reputation and his association with the Aberdeens. The competition was won by Patrick Abercrombie and was the beginning of his life-long association with the city. Dublin Corporation, reeling from the drubbing given to them in the report of the 1913 Housing Inquiry, contracted both Unwin and Geddes to comment and give advice on their housing plans in 1914 (Dublin Corporation report 78/1915).[13] Mears was involved in this analysis and his association with Geddes was further strengthened when he married Geddes' daughter Norah in 1916. Mears confidently expected that he would be asked to develop the preliminary plan for Marino, one of Dublin Corporation's potential locations for suburban housing. Purves' comprehensive study of Mears contains a note to Geddes in February 1915 which stated 'They have brought in a young Dublin man, as a means of greasing the track. He will, I think, be educable and knows the ropes' (Purves, 1987: 85). In the event, the First World War ended any consideration of Marino and Mears was not called upon when Dublin Corporation began the project in earnest in 1918, preferring to rely on their in-house expertise – the 'young Dublin man'. The influence of garden city/suburb ideas can be seen in Dublin Corporation's revised plans for its 'suburb' at Fairbrothers' Fields but most especially in H.T. O'Rourke's designs for Marino in 1919 (see Brady and McManus, 2022).

This did not end Mears' association with Dublin and, in 1920, he was asked by Commander J.F. McCabe, an inspector of the Local Government Board, to visit Dublin and prepare layout plans and perspectives for some new housing schemes. One of these schemes was in Killester, promoted by Henry McLaughlin and E.A. Aston. It was suggested that use be made of a 'cheap and novel' building technique, the patent of which was owned by E.A. Aston (Purves, 1987: 90). Mears was now sufficiently well known in his own right, in Dublin at least, to be referred to by the *Irish Times* as the 'famous town planning expert' (6 October 1921).

Aston also ensured that Mears was chosen as one of the experts to develop the plans of the Greater Dublin Reconstruction Movement. The smoke from the Civil War had barely dissipated when a committee of leading citizens was formed in Dublin in summer 1922 to plan the reconstruction of the city. E.A. Aston was chairman of that committee whose membership included Henry McLaughlin and Lorcan Sherlock, who had been lord mayor between 1912 and 1915.

Although Abercrombie's winning entry in the 1914 competition had yet to be published, the details had been widely known in Dublin almost from the beginning. This committee took the view that Abercrombie's proposals were impracticable, and

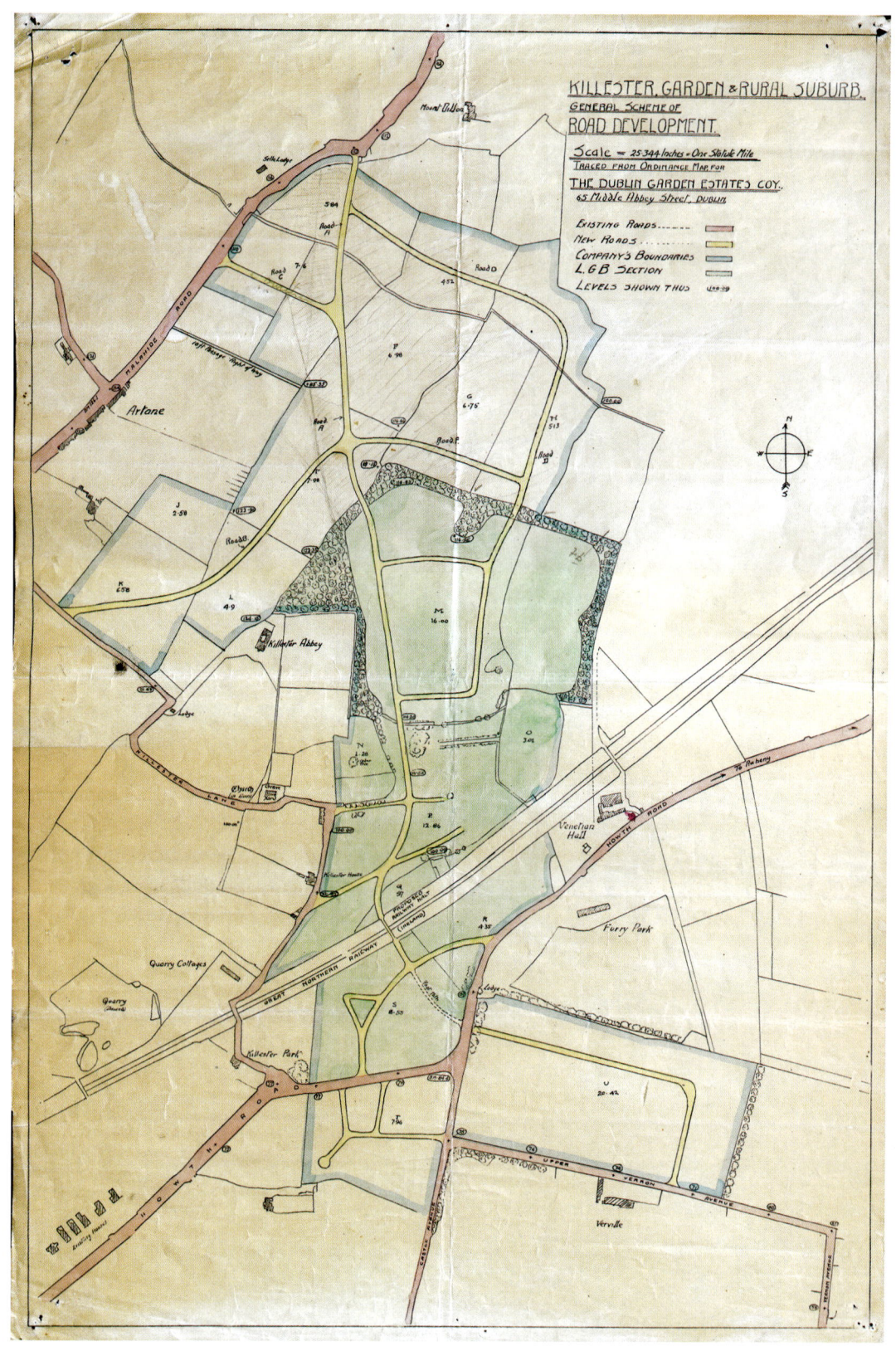

26 Killester, Garden and Rural Suburb. Map of the Killester area showing the land owned by the Dublin Garden Estates Company, outlined in light blue. [OPW/5HC/4/973. NA]

this set up a tetchy relationship between the Civics Institute which would soon publish the plan as *Dublin of the Future* and the Greater Dublin Reconstruction Movement, as the committee came to be called. The debate permeated into personal relationships. It put Mears on the wrong side of H.T. O'Rourke, now the city architect, who championed Abercrombie. It also soured personal relations between Abercrombie and Mears. The public discussion was less confrontational (to begin with at least) and was explained by Aston in an interview with the *Irish Times* which was published on 16 September 1922 (p. 7). He explained that Abercrombie's plan was of 'a suggestive and outline character and to be subject to considerable modification when more information became available. Moreover, while the problems to be solved remained the same, much had changed in the city during the previous seven years'. Mears was the person whom Aston felt best suited to that present task. Any enthusiasm for the Greater Dublin Reconstruction Movement's plans soon evaporated as the cold light of dawn revealed an independent but cash-strapped State. Ultimately, it was Abercrombie who came to be the favoured consultant on planning matters in Dublin until his death in 1957 and Mears had no further official association with the city.

Killester would prove to be one of a small number of housing projects undertaken by Mears; his career took him in other planning and architectural directions. It would be a 'garden suburb' and sometimes it was even referred to as a 'garden city'. Though the label 'garden suburb' was used, it is important to remember that this was not a formal concept – there was no definition of what a garden suburb should look like. It was understood that it would be low density, perhaps with substantial private gardens but certainly with a lot of open space. The layout would make use of the landscape features and avoid any sense of monotony in the layout of the buildings. It was usual to retain existing trees.

To call Killester a garden city would have been somewhat excessive were it not for the fact that a private planned suburb adjacent to Killester was in the air. This was a project of E.A. Aston and it seems clear that Mears produced the design for both the ex-servicemen's houses and Aston's adjacent suburb. The project was endorsed by Henry McLaughlin and it seems that the LGB engineer, F.P. Griffith, also assisted in the preparation of the plans on a personal basis. The same people turned up time and time again in the various aspects of the narrative.

THE PLANS

There is a hand-drawn outline plan of this suburb in the National Archives (Figure 26) which is described as belonging to the 'Dublin Garden Estates Company'. It shows what is described as the 'LGB section' in green with the company's land outlined in blue – not entirely helpful in seeing it clearly. However, the outlining is interesting in that it suggests that the company saw the development as their single entity; with the

Local Government Board (LGB) in charge of one section. This view was reflected in the road network that would connect the various elements. The LGB section separated the private suburb into two components. The larger one lay to the north of the ex-servicemen's houses with a significant road frontage on the Malahide Road and a northern boundary which would have fronted onto what would become Collins Avenue some decades later. The smaller plot was on the other side of the Howth Road to the demesne and took in Castle Avenue and Upper Vernon Avenue. The graphic in the *Dublin Evening Telegraph* for February 1921 brought the overall design to public attention with many of the houses offset from the main roads along more private 'lanes' and surrounding a large village green.

This would not be social housing, it would be privately built, and, for the development of the southern plot, which was within the city boundary, Aston decided on using the newly promoted idea of a public utility society. These had been active in Great Britain since the end of the nineteenth century but their use in Dublin was only beginning. McManus (2004: 2021) has provided the most comprehensive analysis of these organizations in Dublin and the rest of the country. In essence they were co-partnership societies where the members would pool their resources, supported by friends who would buy shares, and build their housing development in stages. Everyone in the society would get a house and the society would ultimately come to an end when all debts had been paid. It was a means of leveraging credit at a time when ordinary individuals found it difficult to obtain loans. Public utility societies would become particularly important to Dublin Corporation as a means of finishing off their developments to a higher standard than they could. Canon David Hall had pioneered the concept of a public utility society in his innovative 'garden suburb' in East Wall and the opening ceremony had taken place very recently on 24 June 1921. The attendance included the local Roman Catholic parish priest, V. Revd Canon Brady, Dr P.C. Cowan and staff of the housing department of the Local Government Board, the city architect and members of the Dublin Corporation's housing committee, as well as representatives of local employers such as the secretary of the Port and Docks Board, the manager of the London and North Western Railway Company, representatives of the Great Northern Railway Company, the Dublin harbour master, and various other dignitaries. The concept of a public utility society had caught the imagination, even if Canon Hall was reported as saying that he found it difficult to raise hard cash (*Irish Builder and Engineer*, 2 July 1921: 449).

The Killester Public Utility Society and the later development on the northern part of the site would add the necessary critical mass to the LGB part scheme to permit the development of a significant range of facilities such as good shops and medical services, perhaps not a town but certainly a suburb. Aston's public utility society was registered on 20 June 1921, even before Canon Hall's official launch. Building had been already underway and the *Irish Times* noted that about ten acres of the possible site had been obtained and between 200 and 300 feet of 'developed building frontage' already existed on Upper Vernon Avenue (*Irish Times*, 28 July 1921: 3). Aston was moving very quickly! The intention was to build about fifty modern semi-detached houses and two storeyed houses, the largest of which would contain seven or eight rooms. At that time, the developing standard for social housing was twelve houses per acre but Aston envisaged having no more than four houses per acre. Unfortunately for Aston, the public utility society was not a success; they claimed that they had difficulty getting access to the subsidies available and they failed to make the necessary returns to the Registrar for Friendly Societies. Despite them claiming that the difficulties had been overcome by 1924, the Register moved to cancel the registration of the society on 23 June 1925. Aston continued with his private building company, the Dublin Garden Estates Company, and they sold sites in nearby Killester and Artane. They complained in 1925, on the eve of the new housing legislation, that they found it almost impossible to sell in the county area because of the absence of grants to private buyers compared to those in the county borough (*Evening Herald*, 25 March 1925: 2). Their activities are a source of confusion because they referred to themselves as operating in the 'Killester Garden Suburb' (*Evening Herald*, 13 March 1925), a term also used for the Killester ex-servicemen's project. Though plots were sold and houses built, this company was not a major commercial success either and it went into voluntary liquidation in January 1937 with a land bank of about forty-one acres in four plots near to the Killester development (*Irish Press*, 16 January 1937: 18).

Aston was not the only one promoting the idea of a garden suburb. In June 1925, the Árd Lorcain Garden Village scheme was described in the newspapers. This was advertised as a middle-class development in Stillorgan (still very much in the countryside) in one of the 'most beautiful and healthful sites possible'. The houses offered were three bedroomed with all modern facilities and big gardens and cost £730 (*Irish Times*, 4 June 1925: 4). However, there seemed little else that would identify this as a 'garden village' and the modest terrace of houses has since been absorbed into the city. The term 'garden suburb' would come into prominence again as building got underway in Mount Merrion. One of Kenny's (the developer) advertisements described Mount Merrion Park and 'Life in Ireland's Garden Suburb'. Ignoring Killester's claims, he wrote that: 'The charm of an historic estate of 250 acres, choicely situated, four miles from Dublin, has been fully preserved, most skilfully planned, and developed into the first Irish Garden Suburb …' (*Irish Times*, 14 March 1936: 8).

Thus, despite the grand designs, it transpired that only the ex-servicemen's element of the suburb came to be built. As mentioned above, the aim of the ex-servicemen's legislation was the repopulation of rural areas and the stimulation of the rural economy. Nowhere was there explicit mention of the provision of urban housing and certainly no mention of how to deal with Dublin or Cork. At the same time, the LGB would have been keenly aware of the needs of Dubliners, having been intimately involved in the development of Dublin Corporation's housing schemes for the working classes. It must have wondered, though, how it could build in the cities and yet be true to the spirit of the legislation, if perhaps not entirely to the detail. It must have been seen as a godsend when the Killester site dropped into their lap.

In fact, the process whereby they got the land was full of twists and turns and, even to this day, it is unclear what happened precisely. In a city where everyone involved in social reform, town planning, house building and business generally knew each other, it is remarkable that there should have been so much confusion and lack of clarity about obtaining the Killester Demesne. Much must have been done on the basis of handshakes and 'understandings' because otherwise there is no explanation for the farrago that ensued.

Sir Henry McLaughlin, chairman of Messrs McLaughlin and Harvey, building contractors, was involved in a variety of activities supporting the war effort. The company was one of the major builders in the city with premises at Dartmouth Building Works, Dartmouth Square. By the 1920s, their advertisements noted that they had built the government offices and the Royal College of Science on Merrion Street, Clery's on Sackville Street, Independent Newspapers, Tylers, Francis Street Market and the play centre at St Patrick's Park. He was hon. director general of recruitment for the British army during the war and also served as chairman of the Local Representative Relief Committee – the local 'branch' of the National Relief Fund. This was a central UK charity whose purpose was to aid the dependents of those on active service and to prevent and relieve 'distress', particularly unemployment, among the civil population arising from the war.

The First World War exposed a lack of food security in the United Kingdom and land across Ireland was turned to agricultural production. Dublin Corporation, in common with other local authorities, acquired land that they put to food production, usually in the form of individual plots or allotments. This was managed by their land cultivation committee and Henry McLaughlin, who represented the Employers Federation, was a valued member. McLaughlin seems to have been highly regarded on this committee and when he resigned in early 1919, the committee wrote in their breviate for the six months ending 31 March 1919 that 'we feel his loss very much' (Dublin Corporation report 144/1919).

With so many men in the trenches, attention turned to training women in the basics of agriculture, particularly in Dublin. In 1916, it suited the Howth estate, owners of Killester Demesne, to offer the land, about thirty-nine acres, to the Local Representative Relief Committee as a training location. According to McLaughlin, the Howth estate was not interested in a lease of less than five years but the relief committee felt that it could commit only to two years. McLaughlin solved the problem by taking the lease personally, with an option to purchase. Or at least, that is how he saw things! No such lease was registered but that does not mean that it was not signed.

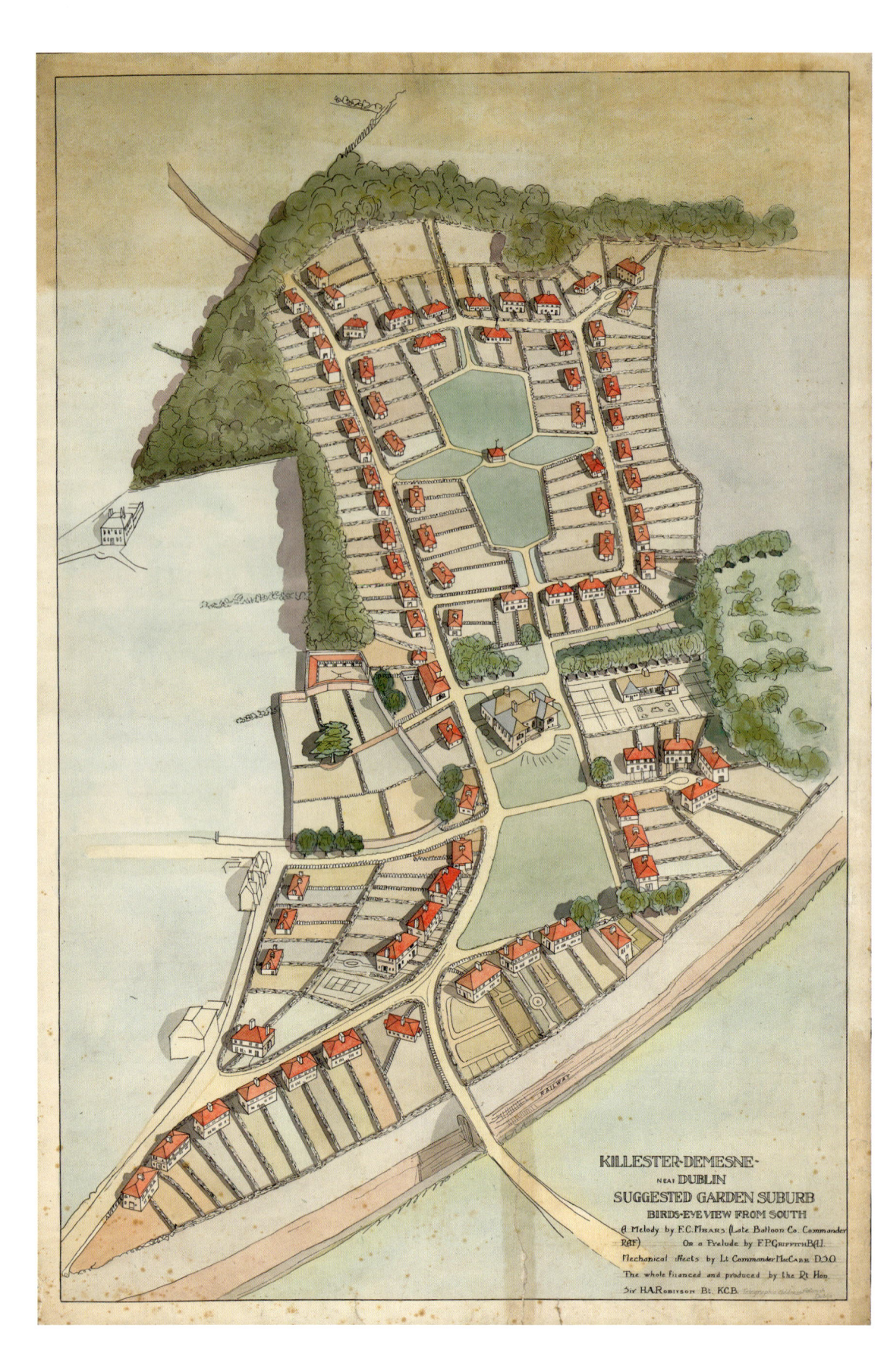

KILLESTER·DEMESNE·
NEAR DUBLIN
SUGGESTED GARDEN SUBURB
BIRDS-EYE VIEW FROM SOUTH
A Melody by F.C.Mears (Late Balloon Co. Commander
RAF)   On a Prelude by F.P.Griffith B.AI.
Mechanical effects by Lt Commander MacCabe D.S.O.
The whole financed and produced by the Rt Hon
Sir H.A.Robinson Bt. K.C.B.

The Killester Garden Village

As the war ended, the relief committee was disbanded by order of the Local Government Board on 31 December 1919 and the question arose as to what to do with the Killester land. It seems that McLaughlin already had a clear view of what to do and had set to developing plans for this section which involved Mears, the Local Government Board and E.A. Aston (as discussed above). There is a beautifully drawn manuscript map (Figure 27) in the National Archives, unfortunately undated, which offers a whimsical view of the team describing the suburb as a 'Melody by F.C. Mears (late Balloon Co. Commander, RAF), On a Prelude by F.P. Griffith B.A.I. Mechanical effects by Lt Commander MacCabe D.S.O. The whole financed by the Rt Hon. Sir H.A. Robinson Bt. K.C.B. Telegraph Address – Astonish Dublin'. The last named was vice-president of the LGB. This has to be an early version of the plan for ex-servicemen and the one McLaughlin would speak about in 1920. It provided for 160 semi-detached houses with a large portion of the site unbuilt. The Killester mansion house was still intact and seems to have been integrated into the scheme. What is also very interesting is that there is a mixture of two-storey and single-storey dwellings. These are similar to those built elsewhere in the country by the LGB. It seems that Mears later changed his mind about the design.

While this planning was underway, the Relief Committee, without McLaughlin knowing, offered the site to Dublin Corporation. This offer went through the formal process and its housing committee wrote to the council recommending 'the acquisition of a plot of ground at Killester, Clontarf, containing about 39 acres, for a sum of £3,500 for housing purposes'. Following some debate, the council voted in favour of the purchase (Dublin Corporation minutes, 6 February 1920). As the council saw things, the lease was for 900 years at an annual rent of £175 with the option to purchase for £3,500 if six months notice was given before the end of the fifth year of the lease. The council was intending to exercise the purchase option. The land was inspected by the housing committee and found suitable for housing. The approval of the housing department of the Local Government Board was obtained as was that of North Dublin Rural District Council because the site lay outside the borough boundary. The lease was obtained and the Corporation entered into legal possession (or so they thought) but since it was not intended to build for some time, the land was then leased to the land cultivation committee on a short-term basis – an eleven-month tenancy (Dublin Corporation report 250/1920). They, in turn, gave it to plotholders, allotments on which individuals could grow food.

It is not as if McLaughlin and Dublin Corporation were strangers. He had been a recent member of the land cultivation committee and his friend and collaborator E.A. Aston was a current active member. Granted, Mears may not have had a close relationship with H.T. O'Rourke so it is possible that the Architect's Office was not aware of the plans being developed. For whatever reasons, the Corporation's plans had progressed significantly by the time that McLaughlin acted. It remains unclear why he

27 Killester Demesne near Dublin. Suggested Garden Suburb. Birds-Eye view from South. [PRIV1232/1. NA]

did not act sooner but what seems to have spurred his reaction was learning that Dublin Corporation did not intend to build on the land for some time. His own project with the Local Government Board was well advanced and would see housing quickly provided for ex-servicemen (see *Irish Times*, 20 January 1921: 6).

He approached Dublin Corporation to sort out the issue with the lease and he met with a sub-committee of the housing committee in November 1920. He told them about the LGB's intention to build 160 cottages and that there were already plans in place for a public utility society to build on an adjacent site of about forty-two acres. In McLaughlin's view, the sub-committee agreed that McLaughlin should have the land and so facilitate the LGB to begin the practical works necessary to develop the site. The view of the housing committee as expressed in report 250/1920 was a bit more nuanced. They felt that there 'were circumstances in connection with the lease which we believe would make it difficult for Sir H. McLaughlin to establish his contention that the lease is his personal property'. However, they were prepared to recommend to the Corporation that they give up the property to McLaughlin because he intended to build there immediately and to do otherwise would simply delay much needed housing. At that moment, the Corporation's scheme at Fairbrothers' Fields was underway, the plans for Marino were in place and early planning had been undertaken for a variety of other projects including Donnycarney, Drumcondra and Cabra.

The matter of the plotholders was also in the process of being sorted out. Independently, McLaughlin wrote to the land cultivation committee and offered £200 to compensate the plotholders who had been given the Killester land to farm. These were members of the Fairview Co-Operative Tillage Society and they seemed happy. This outcome was acceptable to the land cultivation committee; the fact that E.A. Aston was a member would have been helpful. In consequence, that committee wrote on 20 November to the Corporation recommending acceptance of the offer.

This meant that the council had a report from its housing committee recommending that the Corporation step back and let McLaughlin have the land, supported by a letter from its land cultivation committee. It should have gone according to script and taken no more than a couple of minutes. From the general tone of the minutes of the council meeting on 14 December 1920, it seems that members were not entirely imbued with the Christmas spirit. Instead, the council voted not to accept the report of the housing committee and instead endorsed a request that the law agent be asked to prepare a report for a future meeting of a committee of the whole house on the question of the title to the lands. That report was quickly prepared and considered at a meeting on 20 December 1920. It recommended rejection of the housing committee's recommendation for the Killester lands. The report of that meeting went to the council for formal discussion on 10 January 1921 (report 267/1920). Since it had already been discussed at the meeting of the whole house, the business was essentially a reprise of the December debate and the outcome was the same. It was reported in the *Irish Builder and Engineer* that 'at Monday's meeting of the Corporation, a recommendation of a Committee of the Whole House instructing the Law Agent to arrange terms with Sir Henry McLaughlin for handing of the lease and possession of building land at Killester. It was mentioned that there was 30 acres comprised in this site and that the LGB were acquiring it for the building of 160 cottages for the demobilized soldiers. On the motion being put, it was defeated by 22 votes to 5' (*Irish Builder and Engineer*, 15 January 1921: 44).

The *Irish Independent* published a furious letter from the high sherrif, Dr J.C. McWalter, on 12 January 1921 denouncing the decision, which was also published in the issue of the *Irish Builder and Engineer* quoted above. He wrote that: 'For callous brutality, the action of the Dublin Corporation on yesterday stands unmatched' and went on in that vein to say 'the worker is left homeless, hungry, houseless, as far as the Dublin Corporation is concerned. It cannot get the money to build houses itself: it tries to prevent others doing the work' (p. 6).

Other discussions went on behind the scenes but the council agreed to meet in public a deputation from the Irish Federation of Discharged and Demobilised Sailors and Soldiers on 17 January 1921, reported in the *Freeman's Journal*. They were received graciously and, for their part, made the point that they were badly in need of housing and the Corporation's plan would delay that. They asked, as citizens of the city, that the Corporation would deal with the matter in a friendly way. They received friendly support from many councillors but Mrs McGarry indicated that there was some opposition. She was quoted as saying that 'they would deal with them in the friendliest way as citizens. As British soldiers, they would not deal with them at all' (*Freeman's Journal*, 18 January 1921: 6). The lord mayor promised he would hold a special meeting of the council if he could get the necessary support from councillors and this took place on 24 January 1921.

The range of opinions was the same as previously. Mrs McGarry said that she 'regarded the whole thing as a "political swaddle". They had many of their own people in slums, who had not gone out to fight but who were at home concerned for their own country. Those who were manoeuvring this has as object to make them the subjects of charity'. The lord mayor expressed his doubts about McLaughlin's claim to the land while Sir James Gallagher said that his understanding was that the Corporation did not 'own a blade of grass'. John Forrestal wanted to know if the houses were to be given to Irish ex-soldiers or was this to be another 'plantation'. He was also concerned that rents be kept below eight shillings. However, opinions had shifted. Alderman Sir Andrew Beattie had been convinced by what he had heard at the previous meeting and now was co-sponsor of the motion in favour of returning the lands to McLaughlin. These views won out and the recommendation of the housing committee first put to the Corporation in December was now approved (*Irish Independent,* 25 January 1921: 7). McLaughlin was now free to proceed with his intention to make the land available to the LGB. In fact, it seems that the LGB had already taken over the land, even as the Corporation was debating whether to give it to them.

THE KILLESTER SCHEME

Mears endorsed many of the garden city/suburb ideas and suggested a low density suburb for Killester Demesne, which retained many of the existing demesne trees as a boundary. It seems that at this time the demesne was structured into North Field, Middle Field and South Field. Mears kept that basic structure and distributed the 247 houses into three distinct neighbourhoods, Abbeyfield, Middle Third and The Demesne. The Demesne was on the other side of the railway line, giving it a distinctive character. That suited Mears because the scheme would maintain the distinction between different ranks, if the profile of applicants required it. Maintaining a class distinction was not seen as unusual and was factored into the housing development both by the LGB and later by the Sailors' and Soldiers' Land Trust. Mears decided to build bungalows rather than the standard two-storey cottage favoured by the LGB in other developments and this was another characteristic setting Killester apart. The layout of the scheme evolved over the period of its construction and the plans surviving in the National Archives show some of this evolution but, as built, there were thirty-two detached bungalows, each with an area of 1,007 square feet and a parlour, living room, scullery, three bedrooms, larder and coal store. These were intended for higher ranks and mostly located in The Demesne. Another 177 bungalows were semi-detached with an area of 841 square feet, achieved by loss of the parlour or a bedroom. The final thirty-eight had an area of 675 square feet and two bedrooms and the smaller houses in Abbeyfield were largely occupied by army privates. Within the size categories there was variation in the house designs and layouts, as would be expected in a garden suburb. Gardens were provided at front and rear and recreational space was available.

A plan dated August 1923 shows the layout of the water supply superimposed on the final design. This still referred to the North Field, Middle Field and South Field, though it was now a 'Garden Village' and gave the number of each house type within each field. The new names would be given to a scheme of seventy-three houses in The Demesne, forty-nine in Middle Third and 125 in Abbeyfield.

One surviving drawing shows the hand drawn and coloured plan for a Type G2 bungalow at a scale of ¼ inch to 1 foot. This was a three-bedroomed semi-detached house with a living room which was 12 feet by 15 feet. This was a good size and suited to the standard carpet of 12-foot square. The bathroom contained a fixed bath, w.c. and wash basin. Directly beside it was the kitchen which had one built-in press, a sink, larder and a range. There was a door to outside and easy access to the fuel store and a

28 Killester Garden Village Near Dublin. General plan showing proposed water scheme. [OPW/5HC/4/973. NA]

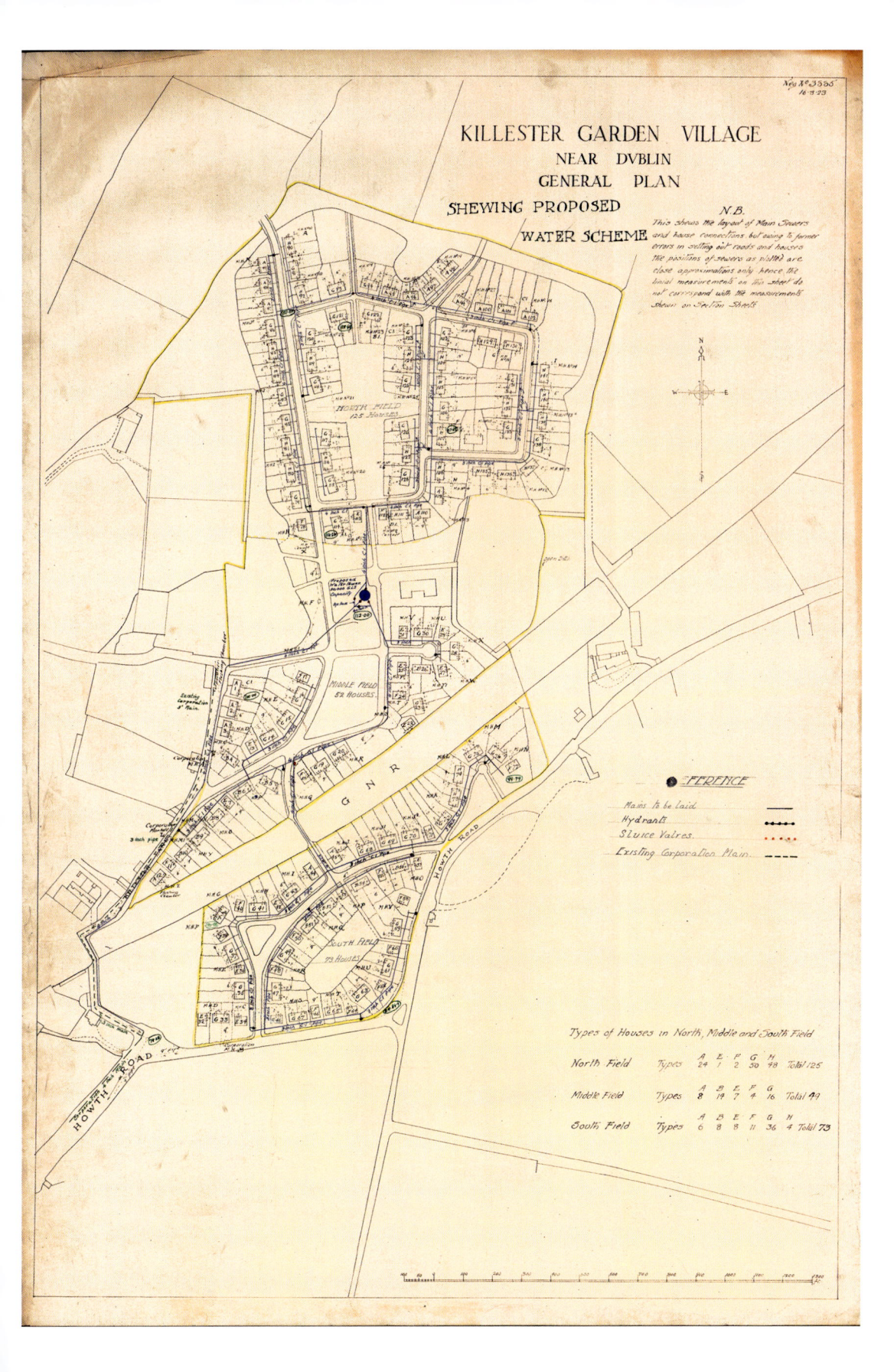

KILLESTER GARDEN VILLAGE
NEAR DUBLIN
GENERAL PLAN
SHEWING PROPOSED WATER SCHEME

N.B. This shews the layout of Main Sewers and house connections but owing to former errors in setting out roads and houses the positions of sewers as plotted are close approximations only, hence the local measurements on this sheet do not correspond with the measurements shewn on Section Sheets.

REFERENCE

Mains to be laid	————
Hydrants	••••••
Sluice Valves	××××××
Existing Corporation Main	- - - -

Types of Houses in North, Middle and South Field

		A	B	E	F	G	H	
North Field	Types	24	1	2	50	48		Total 125
Middle Field	Types	8	19	7	4	16		Total 49
South Field	Types	6	8	8	11	36	4	Total 73

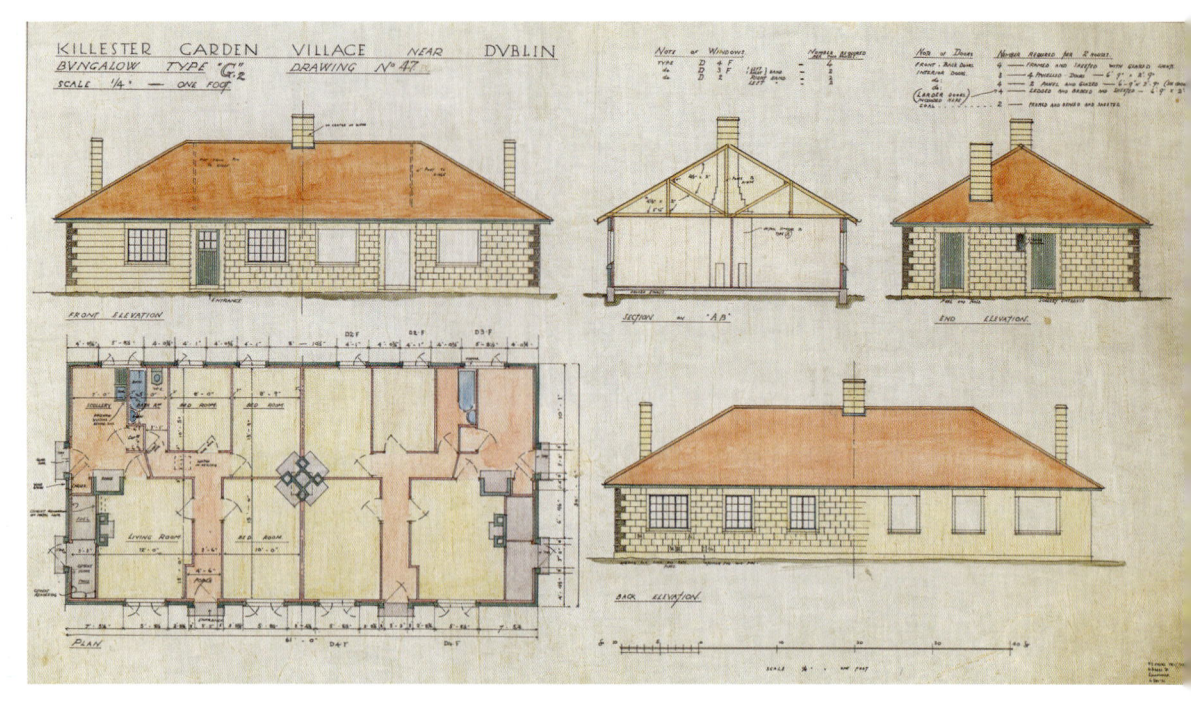

29 Plan for Type G2 bungalow. [OPW/5HC/4/973. NA]

tool shed. A central chimney and a tall chimney to the side resulted in a distinctive roof line. The overall footprint was 61 feet by 30 feet 6 inches. The plan is signed F.C. Mears and dated 6 December 1921 – a day on which another momentous event was taking place.

The Killester estate was of sufficient scale and interest that it was used in advertisements for the Gas Company. The company had undertaken a large expansion of its network and made much of the availability of gas in both Marino and Killester. The advertisement which appeared in the *Irish Times* on 3 June 1926 as part of their Building & Reconstruction series referred to the use of 4½ million cubic feet of gas in Killester during the previous year for cooking, lighting and heating and hot water.

The building process proved expensive and controversial. It was expected that building would be undertaken by the Board of Works, using a building contractor, as happened in other schemes. However, unemployment among the ex-servicemen was high and it was decided that the early phase of construction would be done by direct labour under the supervision of the Local Government Board, in this case the LGB inspector, Commander J.F. McCabe. When the time came to undertake the skilled aspects of the building process, the LGB contracted a local builder who undertook to employ only ex-servicemen. For this, he was paid ten per cent extra for time and materials and it seems that no limit was put on expenditure. Costs spiralled and once

The Killester Garden Village

HM Treasury became aware of what was going on, they cancelled the arrangement, and the Board of Works completed the project using a contractor with a fixed cost contract. The cancelled arrangement resulted in a claim for £49,000 by the builder, H.C. McNally of East Wall, which initially involved a legal dispute but was ultimately sorted out by arbitration. In the statements made during the hearing in the King's Bench Division before the lord chief justice on 5 September 1923 which settled the matter, there was no suggestion of runaway costs. The change in arrangements were explained as being necessary because of 'reorganizations in official circles with the Local Government Board'. Indeed Mr Geoghegan, appearing for the LGB, commented that 'the engineer of the board had very highly commended the work which had been done by Mr McNally, and paid a very high tribute to his services and the efficient manner in which he performed his work. It was only by reason of the reorganization in the office, and necessity for economy, that changes had been made in the carrying on of the work …' (*Irish Builder and Engineer*, 22 September 1923: 721). That analysis would soon change.

The matter was raised in the House of Commons in the select committee on public accounts in 1924 who expressed surprise that no estimate of costs had been made and an open-ended contract entered into with the contractor. This was all the more remarkable given the parsimonious nature of the LGB when dealing with Dublin Corporation's housing schemes. As a result, it was estimated that the houses cost about £1,300 each (later estimates suggested £1,500). It was never made clear exactly what was included in that cost, whether it was an 'all-in' cost or the cost of actual building. In comparison, the houses in Dublin Corporation's scheme in Fairbrothers' Fields, under construction at the same time, cost £659 each to build while those in Marino a couple of years later were built for £550 each, the price reduced as the result of a fierce negotiation with the Department of Local Government. Later, in February 1927, during a Seanad Éireann debate on the Local Government Bill 1926, President Cosgrave stated that the all-cost cost in Fairbrothers' Fields was £950, that in Donnelly's Orchard was £735 while it was £787 in the first phase in Marino and £664 in the Croydon extension. Whichever figure is chosen in comparison, it seems that Killester's costs were way out of line, all the more so when it became clear that they had not been well built. There was no direct relationship between the cost of the Killester housing scheme and the rents charged but they too were higher than might have been expected.

In 1929, it was decided by the Irish Sailors' and Soldiers' Land Trust (see below) to add some forty houses to the scheme, now referred to by the *Irish Times* as a 'garden colony' (10 July 1929: 4). Captain de Lacy, the Trust's inspector, explained to the *Irish Times* that the building would be within some old walled gardens that had not been in use for years. It came as no surprise therefore when this was named Orchard. As the map of the final scheme shows, the Orchard addition is quite different in density and

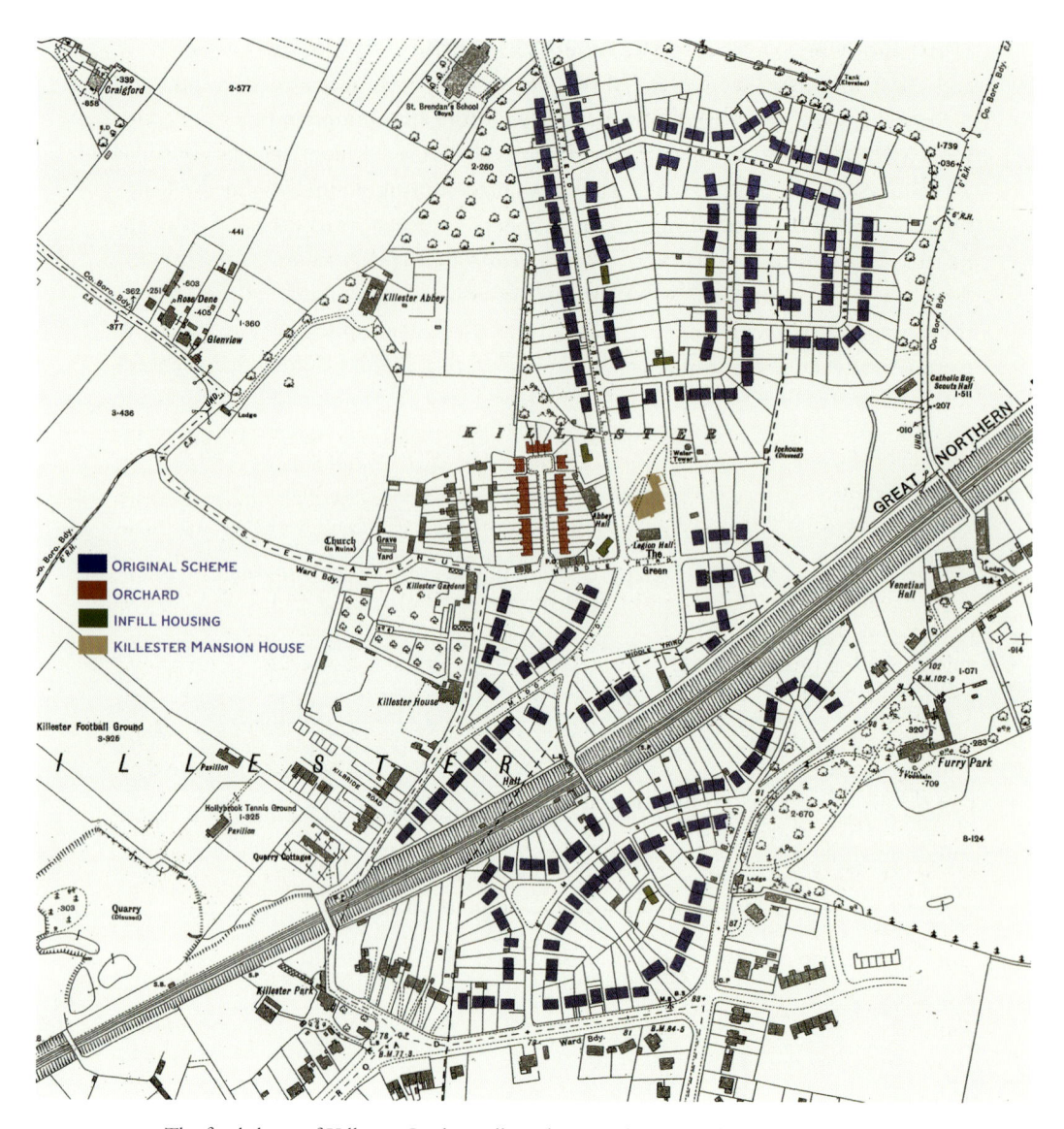

30 The final shape of Killester Garden Village showing the later infill. Ordnance Survey plan, 1:2,500, sheets 15(13) and 19(1), 1938 edition. [PC]

design to the remainder of the scheme. A small number of additional houses were added to the main development and fitted into the scheme by adding an 'a' suffix to the number of the adjacent house. There was originally a cruciform open space in Abbeyfield and eight of these additional houses were added there.

The completed scheme was fitted quite neatly into the landscape with the pre-existing trees on the northern edge largely preserved to give a boundary to the area. To the south, the line of the Howth Road had been maintained to act as a southern boundary, although this resulted in two very awkward traffic junctions. There had been other housing added along the Howth Road and along Castle Avenue and Vernon Avenue, which connected the Killester colony to the rest of the city. However, the impression of a built-up area was misleading as the map shows that farmland was still the dominant use behind the main roads. An open space with the Legion Hall marked where Killester mansion house had stood but other historical elements such as the ice house, the church and the graveyard survived.

THE FREE STATE

Matters changed with Independence. The 1919 legislation suggested that the activities of the LGB in providing housing for ex-servicemen would be time limited and it was never promised that every demand for housing would be met, though that seemed to be assumed by many. As the arrangements for the Free State began to crystallize there was concern that the British government would not continue its support for working-class housing provision in Dublin. E.A. Aston, unsurprisingly, was one of the many voices raised in protest. In one of his letters, reported in the *Irish Builder and Engineer,* he claimed that the 'British Treasury is being allowed quietly to walk off with the Irish Housing Fund'. He asked rhetorically 'What of Ireland, What of Dublin, where housing conditions are infinitely worse than in any English city – where more than one-third of our whole population are crowded five persons to each room in our reeking and crumbling tenement houses?' (30 July 1921: 522). That worry extended to the specific promise to house ex-servicemen and Henry Lefroy was particularly exercised on that front. Whether it was a response to the considerable lobbying that took place, both public and private, or it was going to be done anyway, provision was made for ex-servicemen in the 1922 legislation which separated the Free State from the United Kingdom. Section 3 of the Irish Free State (Consequential Provisions) Act, 1922 set out the mechanism.

> For the purpose of providing in Ireland cottages, with or without plots or gardens, for the accommodation of men who have served in any of His Majesty's naval, military or air forces in the late war, and for other purposes incidental thereto, a body shall be established consisting of five members, of whom three shall be appointed by a Secretary of State, one by the President of the Executive Council of the Irish Free State, and one by the Prime Minister of Northern Ireland (Section 3.1).

This was the genesis of the Irish Sailors' and Soldiers' Land Trust (ISSLT),[14] an imperial Trust, which was given the powers and responsibilities given to the LGB under the

1919 Act. The main omission was that it had no powers of compulsory land acquisition. The act specified that HM Treasury would make up to £1.5m available for the island of Ireland. It was a fixed amount with no mechanism for further payments. Henry Lefroy had previously estimated that 40,000 men might need to be housed, clearly this was not going to come close to this.

Legislation was also required in the Oireachtas and this was dealt with in the Land Trust Powers Act, 1923 introduced into the Dáil on 13 June 1923. It was a short bill which enabled the Trust to function as outlined in the British legislation but it did give the minister for local government the power to acquire land on behalf of the Trust and transfer it to them. There was little debate on the bill in either the Dáil or Seanad. The second stage in the Dáil on 20 June 1923 probably took less than five minutes with Ernest Blythe, the minister for local government, saying 'I think the Bill is not one to which any objection could be taken'. It took a few additional minutes in the Seanad and the bill was signed into law on 16 July 1923. The Trust began work on 1 January 1924 and on 7 February 1924 the order transferring the LGB's land to the Trust was made. Sir Bryan Mahon was appointed by the Free State while the marchioness of Dufferin & Ava was the nominee of the prime minister of Northern Ireland. The secretary of state for home affairs appointed General James Ross, while the secretary of state for the colonies appointed George Herbert Duckworth and Major Harry Lefroy.

If anyone thought that the Trust would be given freedom to act independently of HM Treasury, they were soon brought back to reality. During the time between the passing of the 1922 legislation in the UK and the effective start-up of the Trust, there was a flurry of activity in the British civil service focused on how to manage this expenditure. This resulted in *Statutory rules and orders 1923, no. 1606* entitled *Land settlement, Ireland* and dated 31 December 1923. This set a limit of 2,626 houses in the Free State and 1,046 in Northern Ireland and this included any houses already provided under the 1919 legislation. Including such a level of precision in statutory regulations seems bizarre. However, a reading of the correspondence in the UK National Archives,[15] especially file HO 267/264, shows a laser-like focus in the Treasury on cost and a determination to build for the minimum amount possible. The totals represented what the Treasury believed was possible using their estimates for building cost, site acquisition and site development. There was no suggestion that this would meet the need or anything like that; this was what could be built with the money available.

There is a great deal in that file about housing costs in Belfast because the Treasury could still rely on the co-operation of the civil service and local authorities there and it was assumed that what happened there could be applied to the remainder of the island with some adjustments if necessary. Distinctions were made between housing in rural areas and urban areas and class distinctions were maintained in terms of the housing to be provided to the various categories of ex-servicemen. The Treasury took powers to control how the Trust would spend its money and Regulation 16 allowed them to limit the 'all-in' cost of the house while Regulation 19 allowed them to fix the minimum rent. The Trust pushed back against such management of their affairs and argued that they should be allowed to use their local knowledge to get the best value and to produce the maximum number of houses. This resulted in a detailed inquiry from the Treasury (A.P. Waterfield) in September 1924 (HO 267/264:54) to Stephen Tallents (later Sir) in Belfast asking for data supporting the Treasury's position. The letter illustrates the Treasury's attitude very well. They feared that if the Trust was not given free rein that 'they threaten, quite frankly, to build the prescribed number of houses in such a way that the Treasury will save not a penny on the £1½ million'. To forestall that, the Treasury decided to fix a limit of £500 to cover the 'all-in' cost of each cottage in the North and South. This limit, published on 14 August 1924, was described as 'provisional' in that letter but it was clear that the Treasury felt that they were going to hold to it. What was asked of Tallents was that he produce some costs from Belfast contractors for a standard four-roomed terraced cottage of the simplest possible specification. The costs were to be provided with or without baths because 'I have no doubt that in Belfast we should be bound to provide baths because the standard of civilization there presumably requires it. But I am by no means certain that this is necessary in the South, even in the towns …' The letter went on to ask for the costs of similar cottages in rural districts and 'a slightly better type of cottage, semi-detached, and with an extra room, for the better class of ex-servicemen in the urban areas in case it should be necessary to provide such'. The Trust argued its case strongly in a lengthy memorandum which is dated 8 September 1924. The essence of it was that if they had to build within that limit, they would have to build small inferior housing and 'any proposal to limit their activities to the provision of the smallest type would be to reduce the housing schemes for ex-service men to a dull level of cheap monotony, and would render it impossible to formulate schemes which, either in their amenities or their general aspect, would be a credit either to the Trustees or to the British Government. A certain proportion of better class houses must be built for the warrant officer, the non-commissioned officer and the better type of ex-service man …'

The desire for economy did not end there. The £1.5m was an upper limit but it was now proposed to reduce it. The Treasury took the view that expenditure incurred in the period between the passing of the legislation and the end of 1923 should be charged against that maximum. This came to £350,970 5s. 9d. (the precision is remarkable here too) and this meant that 'a serious effort must be made to complete the work of the Trustees by a capital expenditure of not more than (say) £1,150,000 as from the 1st January 1924' (HO 267/264:80). This, according to the Treasury's calculations, would be more than enough to discharge their obligations, indeed they expected to spend considerably less than this.

The Trust members were not without political influence and negotiations dragged on until March 1925 when the Trust agreed to a full and final settlement of £1.3m but got freedom to act in terms of numbers of houses built, unit costs and rent levels. That freedom to alter the number of houses built was limited by the total amount of money available. The original intention was that the administration of the Trust would be undertaken by the colonial office in Dublin and the ministries of home affairs and finance in Belfast. However, following objections to the costs being attributed to the Trust (and not just from Lefroy), administration in Dublin passed directly to the Trust in 1925. Lefroy had managed to negotiate an extra £50,000 and was praised for his efforts but his experience as a member of the Trust does not seem to have been a happy one. He felt that rents in the Free State were too high and kept high by excessive administrative costs. His active membership on the executive of the British Legion was seen to be problematic by the Trust's chairman and a fundamental conflict of interest. Lefroy was removed from the Trust in 1926 and replaced by Lord Ardee but he continued to be a strong advocate for ex-servicemen, especially via the British Legion and was an important voice at the Lavery inquiry. Lefroy published his version of events and it makes for interesting reading, especially about the attitude of the Treasury (Lefroy, 1927).

LIFE IN KILLESTER

Killester was, by now, complete. It seems that a strong community spirit developed quickly and the newspapers reported on Armistice Day commemorations and other community events. The commemoration of Armistice Day in 1923 involved a full military turn-out of the ex-servicemen and amongst the distinguished attendance was Major Lefroy, who was greeted warmly. In 1924, the men paraded in Killester before taking the train into the war memorial in Amiens Street station. Remembrance days were observed in the years that followed with a contingent from Killester present at national commemorations such as in 1931 (*Irish Times*, 14 November 1931: 10) or November 1936 (*Irish Times*, 3 November 1936: 8).

There was also a social side to community events. In 1925, all children under fourteen years were hosted at a Christmas party held in Clontarf Town Hall. The *Irish Times* reported that over 2,000 people attended the second annual fete and sports day in June 1928 (2 June 1928: 13). What was described as the 'Killester Diamond Follies party' gave two vaudeville entertainments at Molesworth Hall in December 1928. The *Irish Times* commented that 'the players are all associated with the ex-Service men's organization, and have established a reputation for first class work' (10 December 1928: 8).

The British Legion (later Royal British Legion) was founded in 1921 to provide financial and social support to members and veterans of the UK armed forces and it

31 St Brigid's church, Killester, *c*.1943. [ITA Coolock-03. DCLA]

also played an important representative role. Membership in Killester was high and the provision of a Legion Hall provided a focus not only for the activities of the British Legion but also for the community in general. While it was little more than a modest hut used during building works with an annexe, it was described as being comfortable and well-furnished with a piano and billiards table. It was used for all manner of social events and an application was made in 1926 for a drinks licence. At the court hearing about the application, the local garda sergeant from Raheny stated that there was 'a population of about 1,500 in the colony which was fairly peaceable'. While the application was supported by the British Legion, it was opposed by the gardaí and the Land Trust and was refused (*Irish Times*, 10 March 1926: 10).

During the early hours of 23 April 1928, the Hall was burned to the ground. The gardaí found petrol cans in the vicinity and the ferocity of the fire led them to the conclusion that it was arson. Following a successful claim for malicious damage, a new hall was constructed by the local branch of the British Legion and opened by the lord mayor, Alfie Byrne, on 27 August 1932. The social programme resumed. Relations with the British Legion soured somewhat, at least for some, as time went on. In 1955 the secretary of the Land Trust Beneficiaries Association, who lived in Middle Third, wrote to the taoiseach regarding the tenant purchase scheme then in place. The letter claimed *inter alia* and somewhat inaccurately that: 'At no time has it been admitted by the occupants of Trust cottages that the British Legion was competent or authorized by them to speak on their behalf or negotiate for them. There are 289 occupants of cottages in Killester, and there is no Branch of the British Legion here, and though several attempts were made to start one in the past, it has always met with failure' (14 February 1955).

The majority of the residents were Roman Catholic and their number, together with the growth in adjacent areas was sufficient to justify the building of a new parish church. The foundation stone of the church of St Brigid was laid by the archbishop, Dr Byrne, on 5 July 1925 and it had its dedication in October 1926. It was an indication that the area was not well off that economy was emphasized in the design. The *Irish Times* noted that 'in considering this church, it is well to bear in mind that the problem to be solved was the provision of accommodation at a minimum of cost consistent with providing a permanent building of pleasing appearance' (*Irish Times*, 23 September 1926: 4). The church was opened without the apse completed or the internal plastering and furnishing. This reduced the cost from the projected £12,300 for the design to a more manageable £8,900. The contractor was local, J.J. Flanagan from Drumcondra.

Killester was not an agricultural community but it was still quite rural in those early years and connectivity with the city was limited. There had been a railway halt at Killester in the early years of the railway but it closed in 1847. A new station, a little further north of the original, was opened in 1923 but the frequency with which trains stopped there increased only gradually. This was still the era of unregulated public transport and there were many private bus companies in Dublin. Most chose to differentiate themselves from the main company, the Dublin United Tramways Company (DUTC), by having distinctive names – 'Adeline' or 'Angela' or 'Hornet'. It seems though that the service was more akin to the wild west with companies racing each other to get to potential passengers and the 'Pirate' which ran between the city centre and Killester was regarded as an appropriate name by many. The DUTC was given official permission to develop and run bus routes by the Dublin United Tramways (Omnibus Services) Act, 1925 and it gradually put the others out of business although their era did not end completely until legislation in 1944 allowed the DUTC to acquire any company compulsorily. The 'Enterprise' and the 'Paragon' served Marino while the 'S&S' travelled as far as Donnycarney. The Killester development had its own service into town run by the Contemptible Omnibus Company.[16] This was owned by a Mrs Gilbert with a business address of 1 Aungier Street and it seems to have been formed out of a desire to provide a service for the ex-servicemen. However, the company expanded quickly beyond its Killester to city centre route and by 1928 it was running a number of long distance buses from Dublin to provincial centres. The company often figured in court cases arising from the enthusiasm of its bus drivers but perhaps no more often than might be expected for a company of its size. It quite quickly had competition from the DUTC which introduced its own bus service to Killester on 6 July 1925. The number 43 bus ran from Eden Quay to Killester via either the Howth Road or St Lawrence Road. Both

The Killester Garden Village

HEIGH-HO, COME TO THE FARE.

Intending Passenger: "COME ON, BEN HUR. YOU'RE WINNING."

32 The DUTC and Contemptible Omnibus Company racing towards a passenger. *Dublin Opinion*, November 1925: 677. [PC]

services – the 43 and the Contemptible – drove through the Killester estate. While this might have seemed to be a great service, by 1926 there was significant local opposition to this and a demand that the buses keep to the main road. This led to a public inquiry by the Department of Local Government at which one of the residents, Mr McDonald, was quoted as saying that 'the buses of the two companies were running through the place and it was simply a matter of the buses of one company chasing the buses of the other. The Tramways Co. buses had scheduled times but the "Contemptible" buses, he thought, had none' (*Irish Independent*, 29 June 1926: 8). Competition was indeed fierce and the DUTC would later be accused of predatory practices by introducing new services which targeted other companies (*Irish Times*, 14 September 1929: 6). The DUTC significantly reduced its fares on the Killester route in April 1927, which forced an immediate reduction in 'Contemptible' fares, in some cases of up to fifty per cent. The long-term effect of competition was the merger of a number of private companies, including the Contemptible Omnibus Company, into a new entity called the Irish Omnibus Company in December 1929 (*Irish Times*, 14 December 1929: 9). The new company concentrated on provincial routes and tours of Ireland.

The first provisional edition of the Dublin Popular Edition map produced by the Ordnance Survey in 1933 shows that the development was still quite spatially distinct from the remainder of the city with just the narrow corridor of the Howth Road linking it to the city. The new Corporation estates at Marino and Donnycarney were close by and the beginnings of Collins Avenue can be seen but the overall impression is one of countryside. The picture had not changed greatly by the 1948 edition and it

33 A DUTC bus and the developing shopping area in Killester, 1920s. [PC]

34 Advertisement for the Contemptible Bus Company, showing its diversification into regional routes. Dublin's Transport Network, 3d. *Motor News*, October 1928. [PC]

The Killester Garden Village

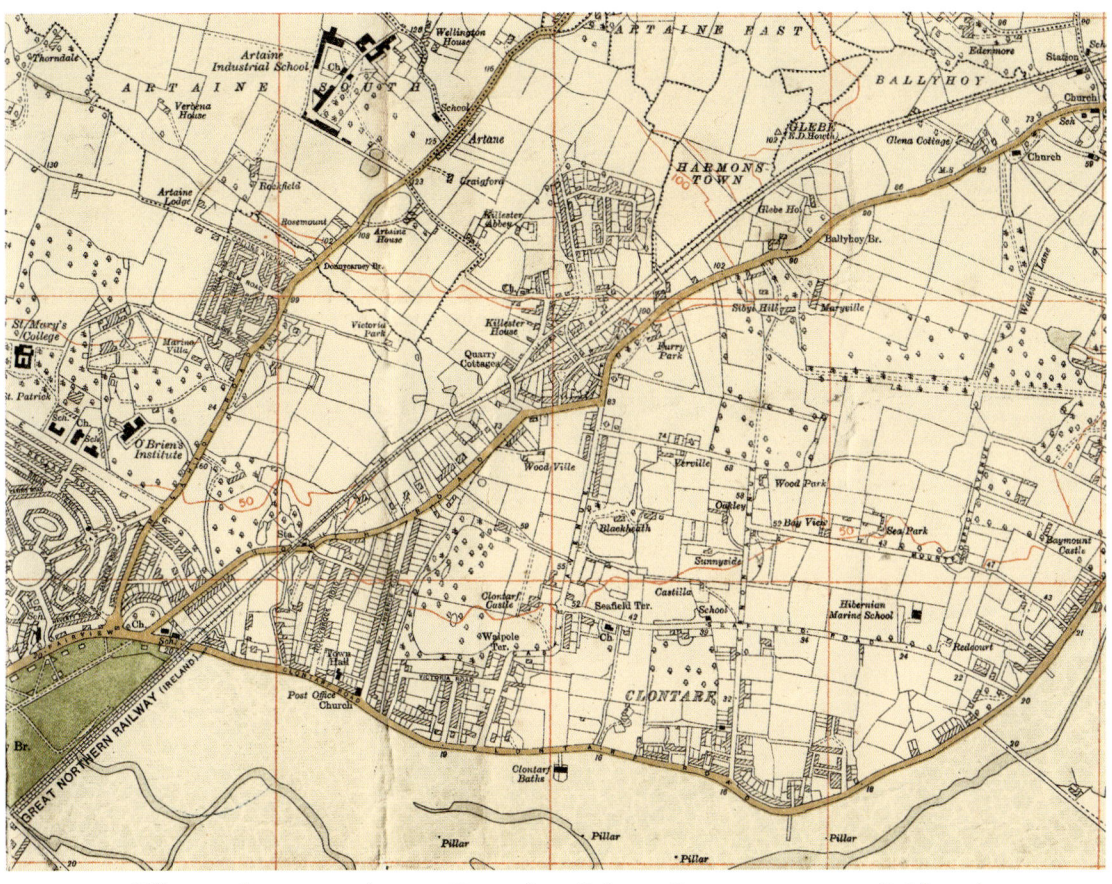

35 Killester and environs, early 1930s. Extract from Ordnance Survey map, 1:20,000, Dublin and Environs, Provisional Popular Edition, Sheet 265b, 1933 edition. [PC]

was only in the 1950s that the city arrived and surrounded the estate. Collins Avenue was extended eastwards to Killester while the Corporation's own developments in Donnycarney and Artane, together with private development in Clontarf served to end any spatial distinctiveness (see the chapter following).

It took a few years for Thom's Directory to list the development but the 1927 edition recognized its distinctiveness and it lists Demesne, Middle Third Street and Abbey Field under a separate heading, Killester Garden City. From 1941, the directories dropped the 'garden city' label but kept the streets together in the sequence – Demesne became Demesne Avenue, Middle Third was promoted to an 'avenue' but Abbeyfield remained as it was. Orchard Road appeared in the normal alphabetical sequence but because the directory still maintained 'Raheny and Killester' as a separate section, it was not far away. A further diminution of distinctiveness was that the houses in The Demesne along Howth Road were integrated into that road and renumbered.

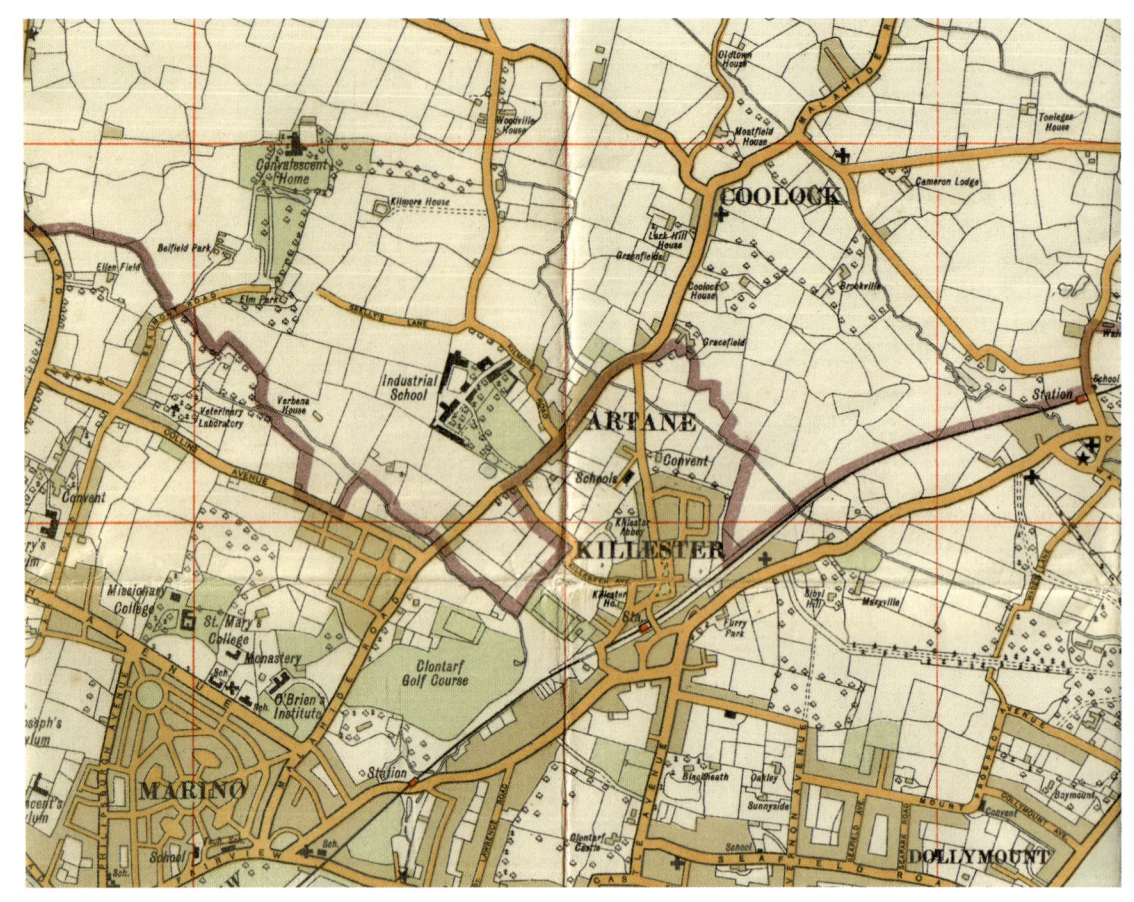

36 Killester and environs, late 1940s. Extract from Ordnance Survey map, 1:25,000, Dublin Popular Edition, 1948 edition. [PC]

It is usual for all residential areas to experience change and even within Land Trust tenancies there was considerable change over the years.

Number of tenants present in each year who were also present in 1926.

	1926	1931	1945	1952
Demesne	73	62	42	33
Middle Third	49	43	34	30
Abbeyfield	125	95	74	69
	247	200	150	132

The figures show that only 132 (53%) of the original tenants (or their widows) were still in residence in 1952. The numbers had begun to decrease quickly and especially in Abbeyfield. Orchard Road, in contrast, was much more stable and twenty-seven of

The Killester Garden Village 77

37 Killester and environs, late 1950s. Extract from Ordnance Survey map, 1:25,000, Dublin Popular Edition, 1959 edition. [PC]

the thirty-two tenants there in 1931 were still there twenty or so years later. A key figure in maintaining continuity in the estate was Capt. James De Lacy. He was appointed as Superintendent of the Killester estate on 1 March 1923, initially with a rent-free house. His duties were to collect the rents and also to act as inspector and liaison officer between the Trust and the tenants. On 1 March 1925, his pay was increased in lieu of his free house and he continued in his role until his death on 25 April 1940. There is a substantial file in the UK National Archives dealing with the Trust's successful request to HM Treasury that his wife be paid a gratuity of £75. In the application the Trust commented on his duties in 'this large and troublesome estate which has always been a source of anxiety to the Trustees'. They also noted that 'the superintendence of the 289 ex-servicemen's cottages on the Trust's estate at Killester has always been a particularly difficult matter calling for special capacity and strength of character. Captain de Lacy exhibited these qualities in full measure during many years of loyal service to the Trust as was indeed to be expected from his fine record of military service in which he had earned a substantive commission and the M.C. with bar as well as the Long Service and G.C. Medals' (T164/188/30).

OTHER SCHEMES

While the process of establishing the Trust was going on, building continued across the country in many schemes already sanctioned by the Local Government Board. These were smaller-scale schemes such as #41 in Navan which comprised twenty-one cottages of which eleven were at Liscartan, three at Ardbraccan, one at Curraghtown, two at Slane, two at Beauparc and two at Dollardstown. They were built according to the Board of Works templates for rural housing, some semi-detached, some on their own. Type II, for example, was two storey with a large living room, scullery and two bedrooms on the upper floor. There could either be a parlour downstairs or a third bedroom. Type III was a semi-detached bungalow with a bedroom in the roof space, two rooms downstairs in addition to a living room (*Irish Times*, 29 November 1923: 4). A 1924 scheme in the Mullingar district comprised sixty-two houses distributed across the environs with only six in the town itself. There were sixteen in

38 Layout of Seafort Gardens. Ordnance Survey plan, 1:2,500, sheets 18(12) and 18(16), 1939 edition. [PC]

The Killester Garden Village

39 Seafort Gardens at completion. [Joseph Brady and G. & T. Crampton photograph archive, UCD]

40 Housing on Fairhill Road Upper, Claddagh, Galway. [AP7/171. TNA]

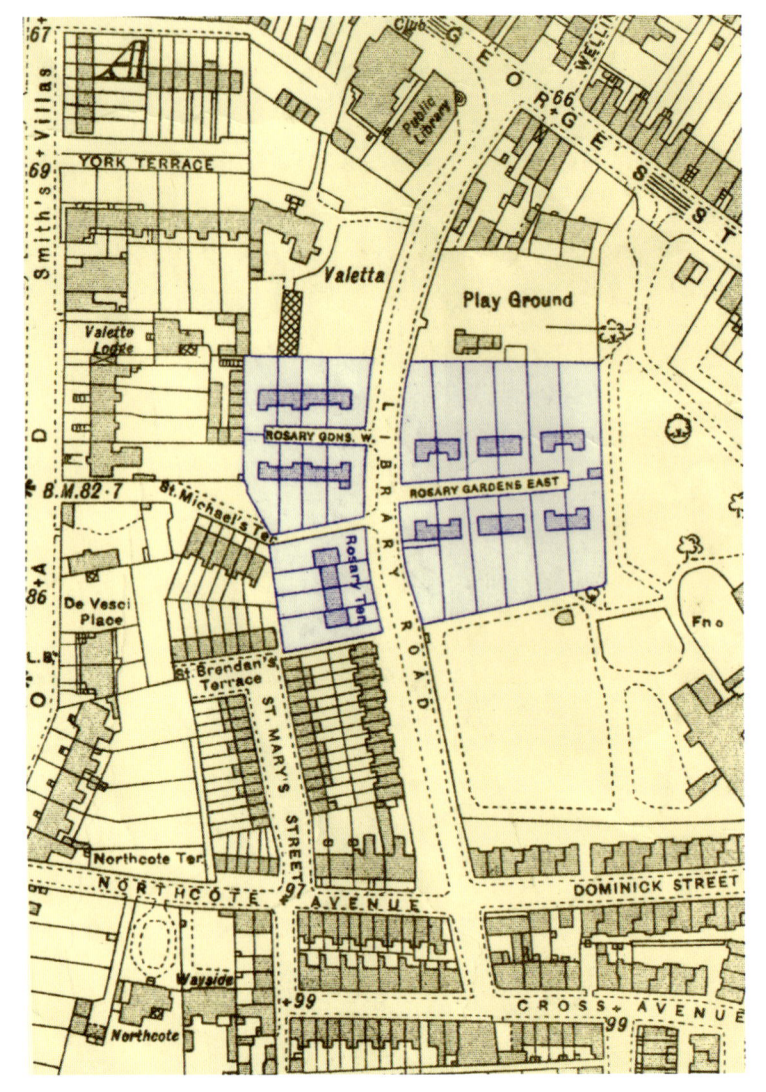

41 Layout of Rosary Gardens, Library Road. Ordnance Survey plan, 1:2,500, sheet 23(6), 1939 edition. [PC]

Clownmore, ten in Newtown, ten in Irishtown and ten in Ballinderry. All were three-bedroomed two-storey cottages (*Irish Times*, 31 July 1924: 5). Even at this point, it was clear that Killester was destined to stand out as a very distinctive scheme.

Seafort Gardens in Pembroke and Library Road in Kingstown were among the schemes inherited by the Trust. Seafort Gardens was built on the old grounds of the Shelbourne football club by the Board of Works, using G.& T. Crampton as contractors. It was distinctive with variation in the rooflines giving character to the scheme. These were two-bedroomed houses, mostly semi-detached with only four detached houses in the scheme of forty-two dwellings. The bedrooms were upstairs with a bathroom and a hot and cold water circulating system. Downstairs was a parlour and a sitting room / kitchenette, together with a coal store, scullery and w.c. Further differentiation in design was achieved by varying the materials used in the roof and their colour. The cost of these houses varied between £593 and £628 and the first of them was ready for occupation in June 1923. The same design was used in Galway on Fairhill Road Upper in the Claddagh for a scheme of forty-six houses, which was completed in June 1924.

In Kingstown, on either side of Library Road, the development comprised twenty-four houses and were built by the Board of Works, using A. Panton Watkinson of St Stephen's Green as the contractor. This company was better known as shop fitters and painting contractors but they seemed to take on building projects too. On the east side, the houses were semi-detached while on the other side of the road, they were in blocks of four. These too were two bedroomed but with a parlour and living room, together with a scullery, coal store and w.c., all on the ground floor. The bathroom was upstairs and there was a hot and cold water circulation system. The design was similarly distinctive but made even more so by the use of red tiles on the roof and coloured blocks on the front walls, shades of light red, white, blue, blue-black and grey. Not so noticeable but important was the use of ten-inch thick cavity walls. No internal plastering was done, the blocks being fashioned so that they were smooth and designed to be papered on directly. This allowed the houses to be built for slightly more than £600, though this was in line with other builders, and they were nearing completion in the last quarter of 1923.

AFTER 1925

As the Trust commenced operations in 1925, it found itself with 1,508 houses completed with a further 112 houses under construction (Trust report 1924–1926). Even with relatively little money, it meant that the Trust had a fairly small target. They set about building immediately and between 1 January 1926 and 31 March 1926, signed contracts for 134 houses. The *Irish Times* reported in August that houses were being built in Athy (8), Cork (70), Drogheda (12), Kingstown (4), Queenstown (16), Sallynoggin (20) and Sligo (20) with schemes soon to commence in Cashel (10), Cahir (8), Cork (24), Fethard (4), Limerick (50), Milltown (36), Sallynoggin (4), Tralee (14), Tullamore (10) and Waterford (30) (7 August 1926: 7). Building continued into the early 1930s and the newspapers regularly had invitations to tender for schemes across the country. For example, there was a tender process for eight houses in Kilrush in February 1929, six houses in Rathdowney in March of the same year while a larger development on thirty houses in Friar's Walk in Cork was underway in May.

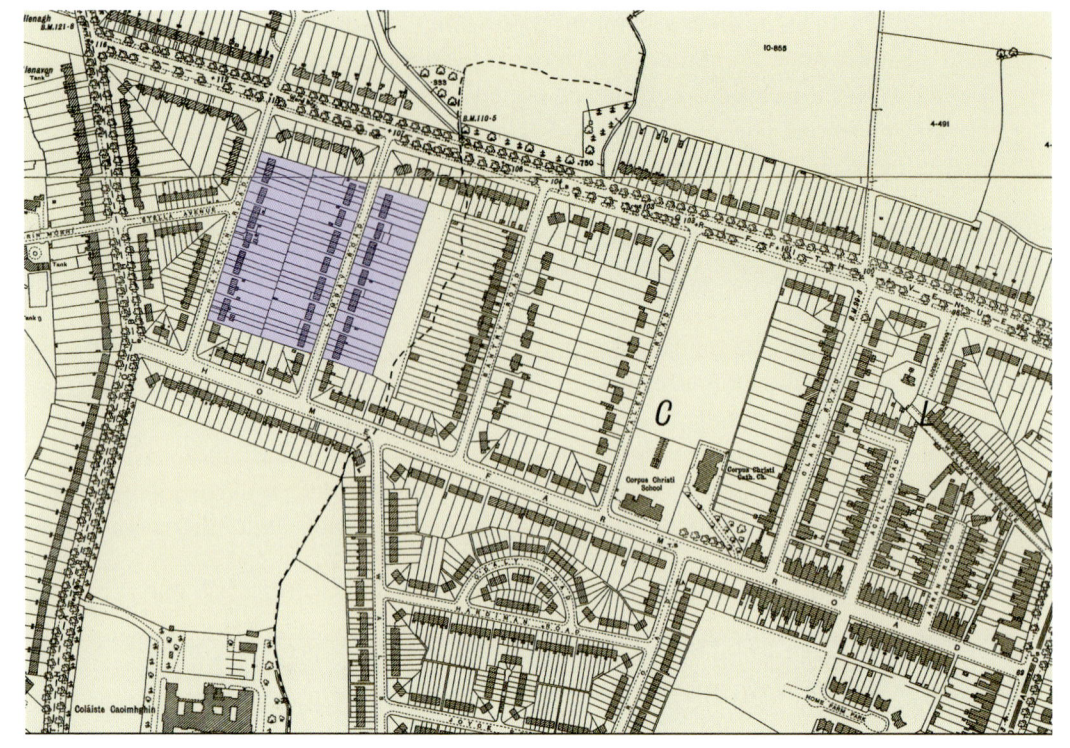

42 Layout of the ISSLT housing scheme in the Drumcondra reserved area. Ordnance Survey plan, 1:2,500, sheets 14(15) and 18(3), 1938 edition. [PC]

The two Dublin developments (Sandymount and Kingstown) mentioned above were outside the Dublin Corporation borough boundary but the Trust also built both within the city and the adjacent townships. The Corporation considered tenders for housing in its Drumcondra scheme in mid-year 1927. As part of that, they also planned an extensive reserved area. Reserved areas were an innovation of the Corporation whereby in order to improve the overall character of a social housing scheme, a portion of the land was made available for private housing, largely but not exclusively built by public utility societies for their members. The idea had been developed at Marino but what was planned for Drumcondra was even more extensive, in fact it might be said that the Corporation lost sight of their main mission (Brady and McManus, 2021).

Dublin Corporation had been prorogued in 1924 for a variety of reasons but mostly because it was not trusted by the government (see Quinlivan, 2021). The city was run by three unelected commissioners until 1930 and they suggested to the ISSLT that they might be interested in building in the Drumcondra reserved area. It is unclear whether this was because of some particular admiration of the work of the Trust by one or more of the commissioners or whether they were asking any organization that might have the capacity to make a contribution to house building.

The Killester Garden Village

43 ISSLT crest on the façade of a house on Lambay Road, Drumcondra. [PC]

The site suggested was at the edge of the reserved area and occupied about 4¼ acres on which the Trust proposed to build forty houses (Dublin Corporation report 194/1927), though they ultimately built sixty-six houses. On the face of it, it was a standard commercial transaction and the price to be paid would be determined by arbitration. However, they were getting the fee simple, the Corporation usually let its land, and the Corporation offered to extend Home Farm Road as far as needed and make two service roads that would link the new development to the new boundary road to the west, what became St Mobhi Road. They were also provided with sewers, water and lighting. Finally, they did not have to pay upfront but rather were allowed to begin building as the details were being worked out – in fact that took until 1938. It was a good (but not excessively so) deal compared to others in the area but the commissioners would have been aware that the Trust did not qualify for the supports available to public utility societies or, indeed, to commercial builders.

Even though there was a desire for maximum economy, the Trust took some trouble with the development and built short terraces, combined with red bricked semi-

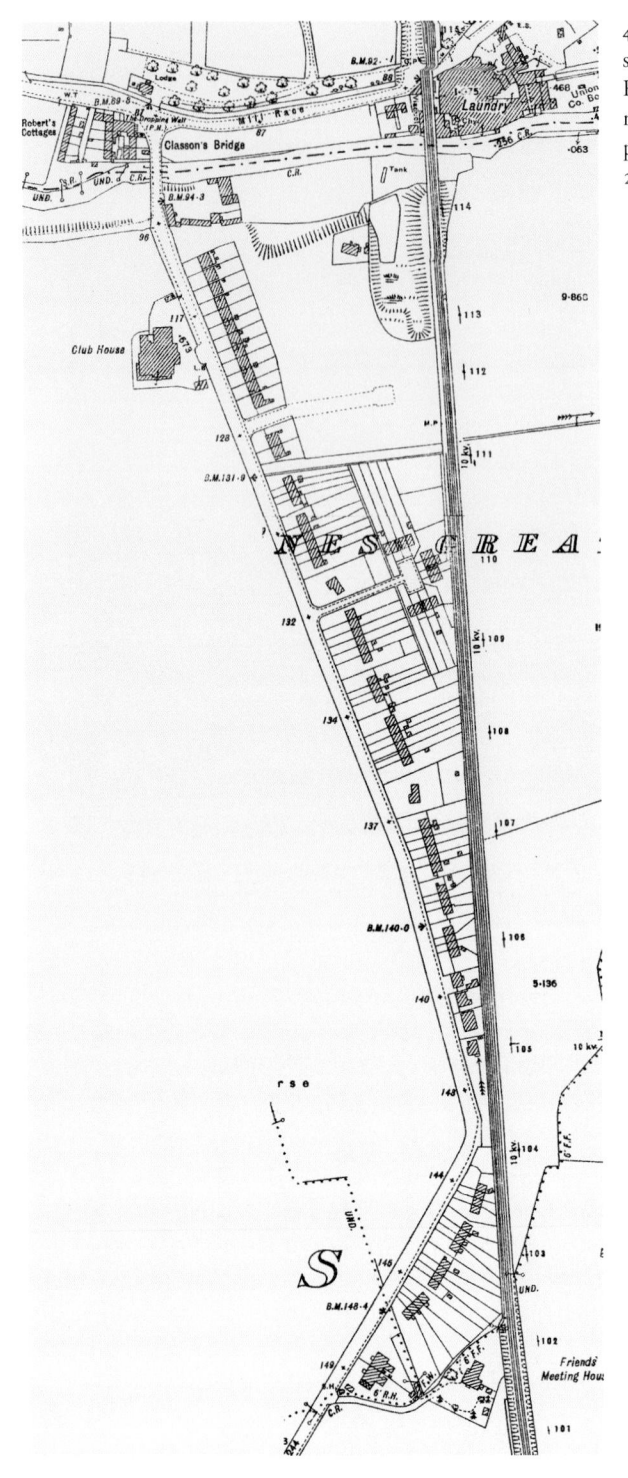

44 Layout of the Milltown scheme between Churchtown Road and the Harcourt Street railway line. Ordnance Survey plan, 1:2,500, sheets 22(3) and 22(7), 1938 edition. [PC]

The Killester Garden Village

45 Layout of the Kimmage (Larkfield) scheme. Ordnance Survey plan, 1:2,500, sheet 22(2), 1938 edition. [PC]

detached houses and used setbacks to avoid monotony in the building line. The houses had three bedrooms, a parlour, sitting room and the other amenities usually present and the average cost in 1927 was about £650. The Trust was never shy about proclaiming its identity and this often took the form of a crest on the façade of one of the houses, in this case (Figure 43) one of the red bricks in the middle of the houses.

In their fourth annual report (1929), the Trust noted that their financial circumstances allowed them to raise their target number of houses to 2,688 in the Free State. The 1930 Housing Act made a grant of £45 available to new builds between July 1930 and April 1931, so a further increase could have been contemplated. In any event, the Trust was getting close to their maximum and their sixth annual report confirmed that 2,299 dwellings were in occupation in the Free State on 31 March 1931.

The larger schemes in Dublin included Milltown and Kimmage. The Milltown scheme was on an awkward site, a narrow strip of land between Churchtown Road Lower and the Harcourt Street railway line. It meant that the Trust could not achieve any sense of self-containment but they built to the design of W.J. Brown mostly in

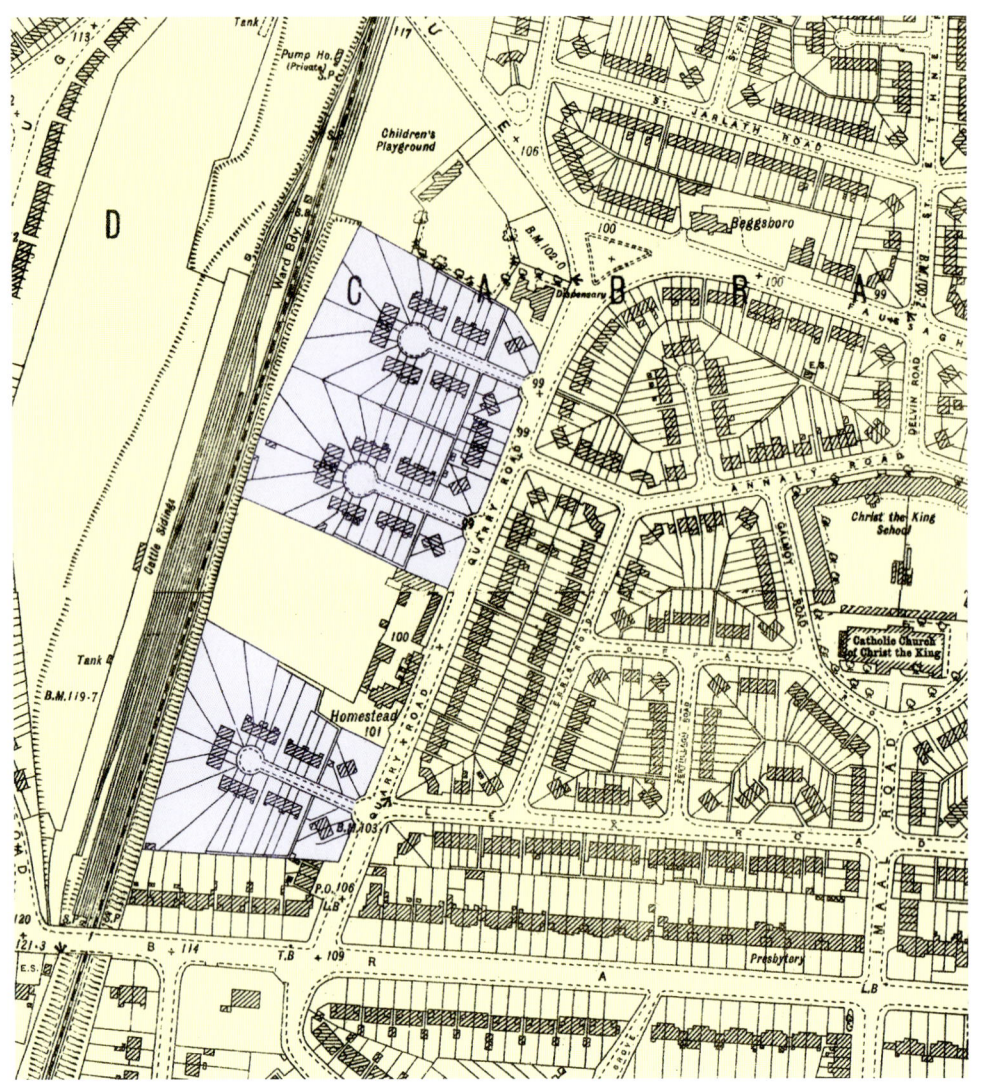

46 ISSLT at Quarry Road, Beggsboro, on the edge of the Dublin Corporation Cabra housing area. Ordnance Survey plan, 1:2,500, sheets 18(2) and 18(6), 1943 edition. [PC]

short terraces of four or six using bricks of different shades and with distinctive fenestration, including faux shutters, and an interesting porch. The contractor was S. Henry & Sons from Drogheda and the forty-eight houses were nearing completion by June 1929. Churchtown Avenue, a narrow road leading to houses arranged around the other sides of a rectangle, has a different design with adjacent doors within a tower-like addition to the façade. The Trust crest appears to be missing from this scheme but the date '1929' appears in a lozenge on a house mid-way along the main road.

The Killester Garden Village 87

47 The entrance piers to one of the culs-de-sac on Quarry Road. [PC]

Another scheme was underway in 1929 in what was then described as Kimmage but which now would be placed in Terenure. A total of sixty-two houses were built with the scheme completed by 1932, after the area had been absorbed into the city, together with the remainder of the Rathmines township. The houses, arranged in short terraces of two and four, are on both sides of Larkfield Park with an extension on the upper part of Larkfield Grove. The houses are of similar design to those in Milltown but with render used for some façades and red brick for others. The Trust's crest can be seen on the semi-detached house which is built at an angle at the apex of a junction of three roads; it forms an introduction to the development. These houses had long gardens and the original concept of a connection to the world of agriculture was maintained in the provision of an allotment area.

The Beggsboro extension to Dublin Corporation's scheme in Cabra provided another opportunity for a Trust development. There was a site between Quarry Road and the railway line of the Great Western Railway Company. This permitted the Trust to build seventy-six houses in three self-contained culs-de-sac with a more distinctive character than at Drumcondra. Each cul-de-sac was provided with an impressive stone entrance and the house at the focus of the road was given an elaborate roof line. The houses, though, were plain terraces. These were spatially distinct from the remainder of Cabra and from each other and it might be argued that it would have been better had the houses been integrated into the scheme.

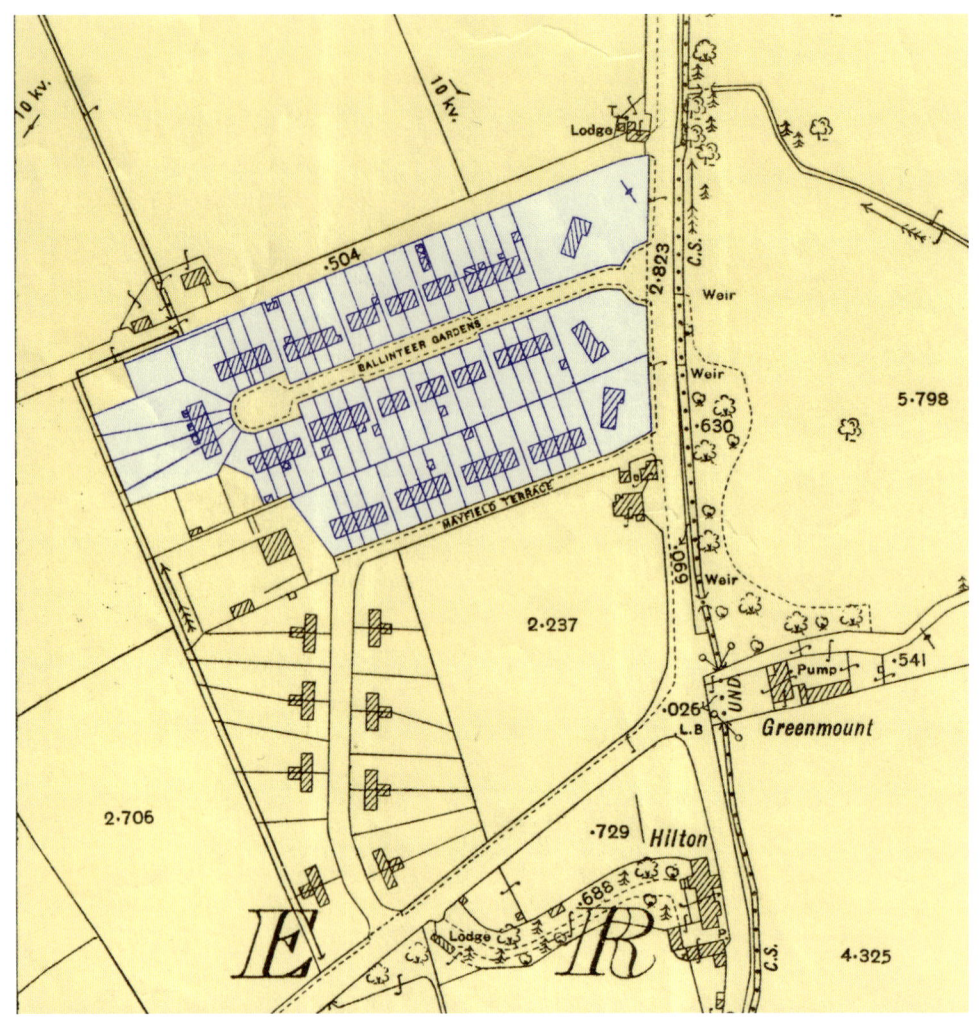

48 The Ballinteer Gardens scheme comprising terraced two-storey houses and bungalows.
Ordnance Survey plan, 1:2,500, sheet 22(12), 1936 edition. [PC]

Another fifty-nine houses were built at Ballinteer Gardens and completed by 1932. This was also laid out as a narrow cul-de-sac but given the distinctive entrance pillars with the added elements of individual entrance pillars for the two large bungalows close to the entrance. This was quite rural at the time and must have shared the same sense of distinctiveness as Killester in those early days. Equally rural and disconnected from other settlement was the development of twenty-eight dwellings in Castleknock. Park Villas was a mixed development of single-storey dormer bungalows and two-storey houses on Peck's Lane. It has long been surrounded by housing and its original character is virtually unrecognizable. Other developments were also undertaken at Sallynoggin, Drumdrum, Raheny, Palmerstown, Clontarf and Sandymount.

The Killester Garden Village

49 The cul-de-sac at Ballinteer Gardens. [AP7/171. TNA]

50 One of the bungalows at the entrance to Ballinteer Gardens. [AP7/171. TNA].

Included in this list was Orchard, the addition to Killester mentioned above. Tenders were sought in October 1929 and it was built as a cul-de-sac with its own stone-pillar entrance, but not as impressive as at Cabra. There was no attempt to integrate the house styles with what had been previously built and the houses comprised short terraces with plain façades.

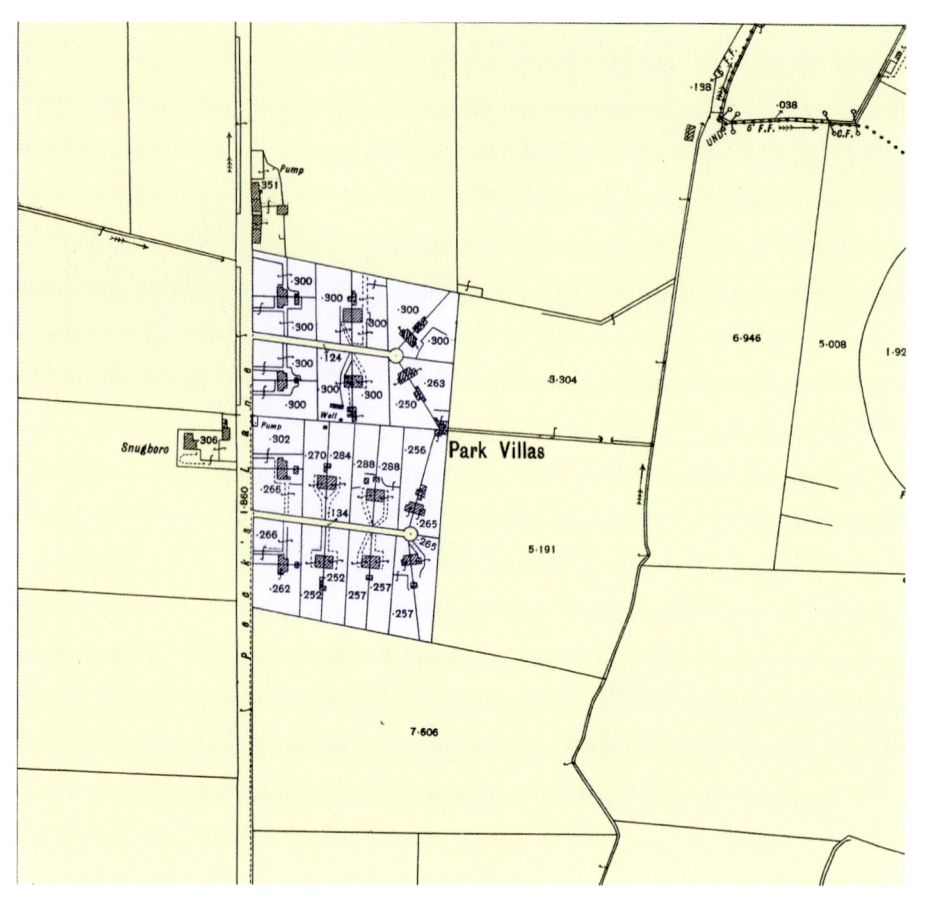

51 Park Villas at Castleknock. Ordnance Survey plan, 1:2,500, sheet 18(1), 1948 edition. [PC]

PROBLEMS IN KILLESTER

Killester was distinctive in spatial terms and also in scale. It was regarded as an expensive place to live and this became apparent with the arrival of Marino and Donnycarney close by. Rents in Killester were 10s., 12s. 6d. and 16s. per week for the different house sizes and for the largest houses this was about the same as the annuity being charged in Fairbrothers' Fields and what would later be charged in Marino and Donnycarney. The significant difference was that these people were buying their houses and were regarded as being at the upper end of the working classes. This was not the case for many of the tenants in Killester and the Trust found themselves in a difficult position from the very beginning because there seemed to have been little consideration of the circumstances of the tenants in the setting of rents by the Treasury. The correspondence in the Trust's archive shows little consideration of local circumstances or the particular circumstances of the tenants anywhere by HM Treasury. Global averages for the UK were always cited as the basis for decision making. The Trust took a more pragmatic view and decided that rents were to reflect the cost of maintaining houses and running the Trust. They would not contribute to a sinking fund or to a capital development programme and any incidental savings could be put to new housing. Maintenance was a problem, though, because it seems that the houses had issues from the very beginning.

Additionally, the Trust began work at a time when there was a growing sense of grievance among ex-servicemen. In the view of a significant number, not only were they not being given the priority they deserved, they were being discriminated against. For a variety of reasons, arrears in the Free State and especially in Killester began to increase quite quickly and this became the first of many differences between the experience of the Trust in Northern Ireland and the Free State. The attitude of the Northern Ireland trustee, the marquess of Dufferin and Ava, suggested a hardline approach. Speaking to the annual general meeting of the Ulster ex-servicemen's association in Belfast on 16 January 1925 he noted that tenants had been chosen with little reference to their suitability and, as a result, arrears had risen to £6,000. As he put it: 'in many cases tenants made no serious attempt to pay their rents, and virtually set at defiance all authority. There were bad cases where tenants had paid no rent for two years'. However, he noted that the Trust had the means to put an end to 'such rent strikes' (*Irish Times*, 17 January 1925: 7). While the agitation against rents was widespread, his comments were particularly directed at Killester. As mentioned above, there was a good community spirit in Killester and a collective sense of grievance had manifested itself in a rent strike in 1924. The British Legion represented the men's concerns in discussions with the Trust and, despite the bullish comments quoted above, a deal was agreed reducing all rents by 4*s*. per week. The Trust was not convinced by the case made but conceded that the rents in Killester were higher than would have been set under the new principles developed by the Trust. This did not settle the matter, though the Trust seemed to believe that they had the agreement of all parties (Trust report 1924–1926). Rather it seems that the people in Killester took the view that this was a provisional settlement, pending a fuller consideration of the rents issue. This view was not accepted by the Trust who made the point that rents had to cover costs and there was no escaping this reality, but the protest continued. This took the form of the tenants paying 'protest' rents of 10*s*., 8*s*. and 5*s*. per week plus an additional 2*s*. per week to deal with arrears. The evidence presented to the Lavery committee (see below) suggested that in 1927 some 220 tenants were paying the protest rents, about ten were complete defaulters and seventeen were paying rents fixed by the Trust (Lavery, 28 January 1928).

The rents issue was a manifestation of dissatisfaction about how ex-servicemen were being treated generally in Irish society and protests were not confined to Killester. There was a perception that civil and public service jobs were being denied to ex-servicemen in the new state and typical of that concern was the report of a meeting of the Dublin branch of the association of ex-service civil servants, which took place in January 1925. It was said at the meeting that the treatment 'meted out to ex-servicemen was the very thing to make men turn Bolsheviks at a time when the country was trying to keep down Bolshevism' (*Irish Times*, 15 January 1925: 9). A lack of progress on suitable public monuments was another cause for annoyance. However, this proved to be a case of 'be careful for what you ask' because this agitation led to the introduction of the Merrion Square (Dublin) Bill, 1927 into the Oireachtas in March 1927. This proposed to take the square as the site for such a memorial. It caused a split among ex-servicemen with Sir Bryan Mahon commenting in the Seanad that Merrion Square 'does not, in any respect, fulfil the requirements of such a memorial, nor do I consider Merrion Square, in any way, a suitable site for a war memorial'. The proposal went away but it was undoubtedly the catalyst which led to the square being obtained as the site for the Roman Catholic cathedral, a gift which the archdiocese did not want but could not refuse.

There was also the question of the scale and nature of housing provision and increasingly there was a demand that an inquiry take place into the operation of the 1919 legislation – that which contained the 'promise'. Also in 1925, a meeting took place in the Shelbourne hotel to consider the formation of a committee to deal with the interests of ex-members of the 'British forces in the twenty-six counties'. This would look at allegations that ex-servicemen had been victimized by the Free State government. Those present included Sir Bryan Mahon, General W.B. Hickie, Sir Henry McLaughlin, Major J.S. Myles, TD and Sir Robert Tate (*Irish Times*, 28 March 1925: 9). Sir Bryan Mahon expressed his personal view that there was no basis to the allegations, especially in relation to the actions of the state, though he conceded that there could be issues in the business world generally. The agitation was sufficiently concerning that the executive council of the state decided on an inquiry into 'the claims of British ex-servicemen' on 29 November 1927 with Cecil Lavery, KC as chairman. The committee met during 1928 and held eighteen meetings. The British Legion submitted a comprehensive report that touched on all aspects of the dissatisfaction that included housing, access to employment, war graves and general issues of equity and discrimination. Lefroy, as vice-president of the British Legion in the Free State, was an important witness. The committee presented its report to the president of the executive council on 8 November 1928. Its general conclusion was that there was no systematic discrimination against ex-servicemen and that 'nothing was brought to our notice to suggest that such ex-Servicemen form a class with grievances and disabilities common to them as a class' (Lavery, 1929: 2). This was not to deny that they had problems but theirs were the same problems that people in their social and economic circumstances had to face. In dealing with housing, the committee was of the view that matters lay between themselves and the Trust and had nothing to do with the state.

The perceived failure of the committee to resolve anything resulted in the formation of a tenants' rights association in Killester that met for the first time in October 1929. At a packed meeting on 28 October a list of grievances was outlined relating to rent, the quality of housing, the number of houses and the treatment of widows. These were general to all tenants of the Trust but there was a view that 'the Killester tenants had been victimized for some ulterior motive and the time had arrived to test the matter' (*Irish Times*, 28 October 1929: 13). Reference was made to the poor quality of local services and that 'the houses were simply dumped in a field without any consideration being given to the essential services'. Though there had been another rent reduction in July 1929, the question increasingly being asked was why should any rent be payable?

The simmering anger was ignited when the Trust moved to evict one of the residents, Mr Butler. He had been a tenant since 1923 but had accumulated significant arrears and by November 1926, he owed £26. The Trust moved to evict him but the matter rumbled on until 1929 when notice to quit was served on Mr Butler on 18 October. By the Spring of 1930, the general issue of the operation of the Trust and the eviction of Butler had been conflated. This can be seen in the letter of 22 March 1930 from Mr D. McAuliffe, the joint honorary secretary of the Killester Tenants' Association to the secretary of state for the colonies (API/135), in which he included and commented on a resolution passed by the association.

> ... we the tenants of Killester, in meeting assembled, condemn in the strongest possible terms the action of the Irish Sailors [*sic*] and Soldiers' Land Trust in their threatened eviction of one of our numbers, Mr. Robert Butler. That we pledge our support to his defence in court when his case comes up for hearing. That we call on the Government of Saorstát Éireann to request the British Government to make an exhaustive inquiry in public into the administration and general working of the Trust, such inquiry to be held in Dublin ...

A neutral observer reading the Trust's archive might come to the view that the Trust was reluctant to confront the ex-servicemen or to evict any. There was an appreciation of their service and their circumstances, but the Trust needed the income from rents to support its activities and the maintenance of the estate. There was also a sense of disappointment because discipline was expected from ex-servicemen. Even though the Trust managed to get an eviction order in May 1930, they offered Butler the opportunity to pay off the arrears at a rate of 3*s*. per week. The Trust also tried to reduce the temperature by holding meetings with the local parish priest, Fr Traynor, and by engaging with the dominions office asking (HO45/14199) about the possibility of rent reductions and additional housing. By then it was too late and a fundamental challenge to the nature of the Trust was in preparation.

KILLESTER AND THE TRUST

On 31 January 1931, proceedings were begun by nine Killester tenants – Robert Leggett, John Connor, Joseph McCann, Richard Healy, John Masterson, Michael Kelly, Peter Markey, Robert George Butler, Hubert McGowan – in a legal process that proved to have fundamental implications for the Trust. The claimants argued that all ex-servicemen occupying Trust cottages in the Free State were entitled to have the cottages vested in them free from obligation to pay rent. They sought an examination of Trust administration, its operating cost structure, and the accumulation of funds to form a reserve fund. They joined the Irish attorney-general in the proceedings on the basis that he represented the State and the public interest, which had failed in their duty of care to honour the promises made during the latter days of the war. The judgment in what became known as the 'Leggett case' was delivered by Mr Justice Johnston in the High Court on 10 March 1932 and the tenants lost (Irish Law Times Reports, 1946).

The case was not about 'the' promise but about whether the legislation permitted, among other things, the Trust to charge rents. Mr Justice Johnston praised the arguments put to him: 'The statement of claim in this case displays much skill and resource on the part of the learned counsel who framed it; but I am bound to say that, having studied its phraseology with the utmost care, it shows a want of clarity – a certain elusiveness that has caused me some perplexity'. However, while being sympathetic to the ex-servicemen he found that:

> But whilst I express my strong approval of the paramount purpose of this great housing scheme for the benefit of those unfortunate ex-sailors and soldiers, and whilst I take leave to say that I entertain a lively hope and expectation that the scope of that excellent work will be extended and enlarged, I cannot say that I have any sympathy whatever with the far-reaching claims that the men have put forward in this case – claims which in my opinion, cannot be substantiated by any of the allegations of law or fact set out in the statement of claim, and which certainly have no basis in the traditions of the housing legislation to which I have briefly referred.

It went to the Supreme Court and the Court upheld most of the judgment of the High Court. However, Mr Justice Murnaghan, who wrote the judgment, found that there was no legislative basis for the Trust acting as they did and that it was not the intention of the legislature that rent should be charged. Interestingly, the same case was pursued in Northern Ireland but there the Court of Appeal upheld the judgment of the Dublin High Court.

The initial euphoria in Killester diminished somewhat once the implications began to be understood. Rent was no longer payable but the extent to which the Trust was liable for repairs and the general maintenance of the estate now came into question with the Trust taking a hard line. There was also a serious problem with continuity. They were life tenants, it seemed, but widows and children had no rights of continuity once the tenant died. A later case, *Casey and others*, which was heard in the High Court in 1936 and the Supreme Court in 1937, clarified the question of tenancy. Mr Justice Murnaghan's written judgment stated: 'The technical legal position of the selected ex-service man is that of tenant at will or tenant by sufferance of the Trust which is charged with the duty of administering the trust of its lands for the benefit of all ex-servicemen. The Trust will no doubt control the property in a reasonable manner and there is nothing to prevent the Trust from permitting a selected ex-service man to continue in occupation so long as he remains a proper object of the Trust'. The path to a solution was outlined in the final sentences of the judgment.

> In reference to some observations that were made by counsel on either side, it is not possible for this Court to offer advice to the Trust as to the general administration of the Trust, nor is it possible for the Court to take upon itself the advocacy of the claims of the ex-service men. I only wish to say that the years are passing – the Great War is over for about eighteen years – and it is much to be desired that a good understanding should be arrived at between the Trust and the ex-service men so that as many ex-service men as possible should enjoy in peace and quiet in their declining years the cottages which the Trust has been able to provide. If, as seems possible, a solution cannot be obtained without the assistance of the Legislature, the Trust and the representatives of the ex-service men must endeavour to hammer out an agreed measure of settlement which might be submitted to the appropriate Legislatures for adoption.

The Trust had, in a moment, lost its income and had to adjust to new circumstances in the Free State. They hoped to negotiate a new arrangement because the position of the tenants was hardly ideal, especially as the Trust declined to undertake maintenance in the years that followed. The tables of income and expenditure from the accounts for 1935 show this starkly. They noted that rent had not been charged in the Free State since the finalization of the case and the data showed a rental income from there of just under £78 compared to nearly £16,000 in Northern Ireland where rent continued to be paid. This was reflected in the expenditure on maintenance which had fallen in the Free State to just over £1,300 compared to over £5,100 in Northern Ireland.

With a view to finding a solution there were discussions with the de Valera government, but that government played its cards very carefully. It allowed discussions to continue interminably and then set up a committee to look at the matter further. The *Committee of inquiry into property administered by the Irish Sailors' and Soldiers' Land Trust* was appointed by the government on 6 March 1940 and held three public sessions in April of that year. Their terms of reference asked them to 'examine the nature and extent of the difficulties which exist in relation to the property administered by the Sailors' and Soldiers' Land Trust, and to make recommendations as to the remedial action calculated to remove such difficulties as exist without imposing any charge on State funds'. Among the specifics of the issue, they were to deal with the issue as to whether widows should be permitted to remain in occupation and the conditions under which this might happen. They took evidence from the Trust and from the Ex-servicemen's Tenants' Rights Association as well as from tenants in Killester and Cabra, the latter being on behalf of the Irish Sailors' and Soldiers' Land Trust Tenants (Cabra area) Association. These witnesses told the committee that they had maintained their houses as well as they could inside and out. However, they had found it impossible to engage with the Trust about developing a reasonable approach to minor and major maintenance. They felt that the financial arrangements being proposed for the Trust were unreasonable and beyond what would be paid in similar developments. They also suggested a purchase scheme to provide the Trust with additional income (NLI, MS 17,193/16). Mr P. Griffin of Middle Third put forward a detailed plan for tenant purchase in the event of the fee simple not being granted to tenants. It was unclear whether he was writing on his own behalf or on behalf of the Old Killester Tenants Rights (*sic*) Association and the Ex-servicemen Tenants Rights (*sic*) Association, whose secretary he had been for eight years (NLI, MS 17,193/8). The suggestion from Mr Patrick Walsh on his own behalf and a number of other Killester residents was simpler. He pointed out that as tenants-at-will they did not have the full suite of rights available to tenant renters. They suggested that they become tenant renters with an annual payment of one shilling (NLI, MS 17,193/15).

Notwithstanding the above, it was a rather narrow consultation but the committee professed itself satisfied that it had obtained all that it needed to come to a view. It was a five-person committee with Judge W.G. Shannon in the chair and it included Major General W.B. Hickie and Thomas Johnson, a prominent labour activist and senior member of Fianna Fáil. A draft report was ready by May 1940 for the minister for local government and public health. That report recommended that the Trust spend the necessary money to deal with maintenance issues without charging the tenants. They would also do what was needed to build more houses to meet demand. Tenants were to pay rent based on the continuing maintenance needs of the housing. They would be tenants for life and the Trust could, at its discretion, continue widows in occupation. However, in exercising that discretion, the Trust could not take demand from other servicemen for housing into consideration. The Trust would have power to sell houses for cash to tenants or their widows but only at a price approved by the minister for local government. However, in the committee's view there was no demand for a purchase scheme at that time, except at prices so low as to be unacceptable. Their view was that introducing such a scheme would be to invite abuses and unnecessary cost.

There is no record of a further version of the report and it does not seem to have ever been published. But for the copies which are contained in the Johnson papers in the National Library, there might not be an accessible record of the deliberations of this committee (NLI, MS 17,193/9). When asked over the next few years what had been done to implement the recommendations, the government answered that consultations with the various interests were continuing. Reading the correspondence of the Trust, it seemed that they believed that the Irish government wanted a resolution and was working towards it, even when all the evidence was to the contrary.

By the end of the 1940s, the Trust had finally come to the conclusion that no Irish government was ever going to legislate to deal with the issues and decided that its only option was an orderly winding down of its activities in Ireland (Éire). This was despite significant pressure from the British Legion, among others, to reanimate the Trust in the light of the needs of Second World War veterans. The wind-down meant the sale of its properties, something which had been considered by the Trust from the very early days but for which they had never sought approval. A short act was approved by the British parliament in 1952, which gave the Trust the power to offer its cottages for sale to its occupants. It provided for the sale '(a) to any of the men for whose benefit the Trust was established, including a man in occupation of the cottage to be sold or of some other cottage so provided; or (b) to the widow of any such man dying before or within six months after the commencement of this Act, if (i) they were residing in the cottage together at the time of his death, and she has remained in occupation of it since; and (ii) he or she has, within the said six months, given the Trust notice of a desire to buy the cottage'. The briefing document for the legislation provided a short statement of the position of the Trust at that time (T233/1292). Some 2,720 cottages had been built in Ireland with a further 1,313 in Northern Ireland. In Ireland there were 624 widow tenants. It was noted that: 'the trust in Northern Ireland has had a successful and, on the whole, peaceful career with a contented tenantry'.

The second reading of the bill took place in the House of Lords on 24 June 1952. It was a short debate but useful in that the marquess of Salisbury set out the position of the British government. In the absence of the ability to levy rent and even with minimal maintenance, the Trust was almost out of money. There was no suggestion that more money would be forthcoming and a programme of sale to tenants would have the effect of releasing a significant capital sum to the Trust and the legislation would facilitate a 'sell and build' process. The sum raised would be used primarily to build more houses for those still on the waiting list. It was anticipated that the purchase facility would be of most interest to those in Ireland but the facility would be made available to Northern Ireland too. However, Lord Carew, one of the trustees, gave a somewhat more nuanced account of how the money would be used. The sales would permit 'the Trust to build further houses where specially required in the city areas, to continue the maintenance of present houses and to provide, where practicable, modern water supply and sanitation in conjunction with the recent extension of such facilities by local authorities'. He made the point also that the 'number of houses remaining under the Trust control will materially decrease, thus reducing progressively the burdens on Trust funds of administration and maintenance'. The bill passed without difficulty and it was reported that the Irish government was satisfied with the outcome. This allowed the British government to 'wash their hands and dispose of yet another Irish problem' (T233/1291, 5 July 1952).

It took some time and some hiccups before the terms of the sale were finally agreed but by late 1958 it was reported that 580 cottages had been sold in Ireland (DO 35/9150). Completion of the sales was slow because there were difficulties with obtaining mortgages and there were various legal actions that had to run their course. The Trust did not expect all tenants to be in a position to buy their properties and it committed to retaining rent-free occupation, for widows in particular, as long as necessary. Further amendments were made to the regulations in 1963 with new legislation in 1967. This allowed the Trust to avoid having to build themselves but instead to provide a grant to an organization providing old folks' housing in return for housing ex-servicemen. The 1967 legislation simplified the position regarding widows – they needed to be in occupation at the time of the husband's death and to have remained in occupation – but also simplified the business of determining a price. Up to this point the Trust had specified a 'reduced price' with a stipulation that the house could not be sold on for five years. The scheme now offered the house at the market value less ten per cent with no requirement to retain possession. This was far less generous than the scheme provided for Dublin Corporation tenants where the discount could be as large as thirty per cent for those with long tenancies (Brady, 2016). Nonetheless, this proved acceptable in Ireland but caused a furore in Northern Ireland resulting in the separation of the process into two schemes. The gradual wind-down of the Trust continued until its final end in 1988 with the enactment of the Irish Sailors' and Soldiers' Land Trust Act, 1987 in the UK and a similar act in the Oireachtas in 1988. It had made an important contribution to the housing crisis in Dublin, where slightly over one third of the Trust's output was located and it had created some distinctive landscapes, especially in Killester.

CHAPTER THREE

The suburbanization of Killester in the twentieth century

RUTH McMANUS

Killester was still a predominantly rural area by the time that the Free State came into existence. Among the small population of the area, many were agricultural labourers, as is seen in the census returns from the early twentieth century. In the summer of 1944, old meadow hay was still for sale at Killester (*Irish Independent*, 29 July 1944). The gradual transformation of Killester from being part of Dublin's rural hinterland into a modern residential suburb took place over the course of more than fifty years, and it continues to evolve. While these changes were gradual and incremental, they did not happen at a steady pace. Three main phases or waves of development can be identified, each of which saw a significant growth in population and residential units in Killester. The first of these phases was the construction of housing for ex-servicemen in the 1920s, the second involved a significant local authority housing scheme in the late 1940s and the third phase involved infill housing from the 1970s, including the first apartment development in the area.

KILLESTER BEFORE SECOND WORLD WAR

Although the environs of Killester were generally rural at the time that the ex-servicemen's housing was built, they were not unchanging. Some private development was being undertaken, particularly along the main arteries out of the city, the Malahide Road and the Howth Road. This construction activity increased during the 1930s, when new housing was developed immediately adjacent to the servicemen's housing at Kilbride Road (developed under two building leases in 1934 and 1936 respectively). Indeed, 190 private dwellings were initially planned here, as discussed below. In 1936, builder Joseph Flynn was selling two-storey, three-bed red-brick houses with slated roofs on Kilbride Road from £625. They qualified for Dublin Corporation loans, which were repayable from 13s. 9d., weekly (*Irish Times*, 18 April 1936). The general increase in density taking place in the locality in the late 1920s and early 1930s, and the opportunities presented as the city expanded, are evidenced in the sale particulars of the house known as Killester Park, Howth Road, in November 1930. The *Irish Times* (1 November 1930) reported that the 'very interesting property' was to be disposed of in three lots. The residence itself still stood on about two acres of land and was in perfect condition following a complete renovation. There were three reception rooms, four bedrooms, servants' quarters, electric light throughout, good garden, garage, cow-shed and two greenhouses. The combination of out-buildings reflects continued agricultural and market gardening activities, together with a move towards modern motorized transport. A second lot comprised the ground rent out of six plots of ground on the Howth Road on which five houses had been completed, while another was in course of erection. The final lot was a lock-up shop let at £150 per annum, described as being in a good position at the corner, 'where all the buses stop' and let to a solvent tenant. Clearly a portion of the former lands of Killester Park had already been built upon, with new private dwellings and a shop unit catering to the growing population of the area.

East of the Malahide Road and south of Killester Avenue, the forty-five acres surrounding Artane House were used for both dairying and store cattle, sometimes holding up to 200 animals. However, by 1934, the occupant, Hugh Cleneghan, had diversified into house construction. His 'Artane House Building Estate' at Maypark comprised ten houses and a shop by 1937, although he had hoped to develop the entire forty-five-acre site for housing (*Irish Times*, 18 June 1937). On the other side of Killester, other large houses with grounds also began to give way to housing during the 1930s. New housing estates were being built close to the Howth Road. For example, by 1936, land around Furry Park House was being developed by Thomas Caffrey and Joseph McGonagle (including Dunluce Road). This was one of the larger schemes, with 150 houses planned. Further north towards Raheny village, Ennafort Park was also underway. Following a hiatus during the Second World War, when materials and labour shortages reduced scope for building, house construction resumed. By 1948, builder Joseph McGonagle was advertising newly built four-bedroom semi-detached houses on Dunseverick Road, which had replaced the former Woodville House.

By the time of the 1936 census, it could be expected that the house building undertaken in the Killester area since the previous enumeration ten years previously would be reflected in population growth. Unfortunately, because of boundary changes this is not as easily captured as might be hoped. Whereas the 1926 data had included 'Killester Town' within the Dublin North Rural District, the new boundaries used for 1936, which also abandoned the 'Killester Town' designation, saw the Killester area largely divided between the two wards of Clontarf East and West. The boundaries of Clontarf West remained relatively stable, with only minor changes, but the extent of Clontarf East was increased from 772 acres in 1926 to 1,030 acres in 1936 and 2,042 acres in 1946. Thus, the census results over the decades are not directly comparable. However, by applying the 1936 boundaries to the 1926 data, we can see that the population of Clontarf East grew by almost two-thirds, to 7,051, in the ten years to 1936, while Clontarf West's 106.9% increase largely reflects the growing population of young families in Marino (see previous chapter).

AN EVOLVING COMMUNITY

The development of the scheme for ex-servicemen resulted in a leap in population numbers in the area, which in turn required additional services. The new St Brigid's church on the Howth Road was discussed in the previous chapter. Its architect, J.J. Robinsion, also designed the national school of the same name, which opened in 1928. It was in the care of the Holy Faith sisters whose convent opened in 1932. Initally catering for both boys and girls, a separate two-teacher boys' national school was subsequently provided. The present St Brigid's boys school was opened opposite the church on Howth Road in 1974.

Even after the development of the estate for ex-servicemen, some elements of the former Killester Demesne remained. These included the ice house, which lay under the shade created by trees in a part of the estate that the veterans living in the new houses called 'The Woods'. The woods formed part of a tree walk that encircled the new housing area, a survival from the old demesne. Children growing up in the locality enjoyed playing in and exploring these new surroundings. However, most of the trees were used for firewood during the Second World War. The Woods area is now the site of Haddon Park football ground.

The interaction between existing agricultural interests and the residents of the new housing was not always easy. A district court case was taken in 1935 following damage to hay by 'eight little boys', who had knocked down thirty cocks of hay while playing 'follow the leader' in Mr Walsh's field (Walsh farmed at Venetian Hall). Five men had to be employed to rebuild the haycocks. Miss Early, solicitor for the latter, stated that he 'received terrible persecution from the boys of the neighbourhood and was often struck by stones.' The judge took a dim view of the children's behaviour, fining their parents five shillings each, plus five shillings compensation and five shillings costs, and warning that any future such behaviour could result in a prison sentence (*Evening Herald*, 7 August 1935). Another court case in 1940 saw the Probation Act applied in the case of two boys who had tried to break into Patrick Murray's orchard at Killester House. They were fined twenty shillings. and ordered to pay five shillings each in compensation to Mr Murray, having caused 'malicious damage' to trellis work (*Irish Press*, 27 August 1940).

A SECOND MAJOR WAVE OF DEVELOPMENT: PLANNING 'NEW' DONNYCARNEY

While the pace of private housing development in the vicinity of Killester was relatively slow and incremental in the period following the completion of the ex-servicemen's housing, a more rapid and dramatic transformation was envisaged by Dublin Corporation when they planned a large new social housing scheme in the locality. Already mentioned in Corporation reports in the early 1930s, the proposed 'new' Donnycarney scheme (to distinguish it from the earlier scheme built 1929–31) was in active development from the late 1930s, when lands stretching from the Malahide Road to the railway line and south to the Clontarf golf course were acquired by compulsory purchase. Local disquiet was voiced at the inquiry into the compulsory purchase order, including by Hugh Cleneghan, whose smaller-scale private building activity described above was set to lose out under the new arrangement. Class bias was also evident, as residents of recently built housing claimed compensation for the depreciation of the value of their homes, on the basis that the new local authority housing would 'be an inexpensive type of house let at a low rent' (*Irish Independent*, 23 August 1938).

Nevertheless, the need for housing was pressing, due to tenement over-crowding in the city centre combined with an ongoing flow of population from rural areas into Dublin. In 1938, when Dublin Corporation unveiled a five-year plan to construct 12,000 dwellings at a cost of almost £7.5 million, it calculated that 17,000 families in unfit basements, tenements or cottages and in overcrowded conditions needed to be rehoused (Dublin Corporation report 5/1938). Large-scale suburban schemes were already underway at Ellenfield, Terenure, Harold's Cross and Crumlin. It was envisaged that this new housing development at Donnycarney would provide 1,000 dwellings. The previous Dublin Corporation scheme, 'old' Donnycarney, was located west of the Malahide Road and of a smaller size (Brady and McManus, 2021).

The significant scale of the 'new' Donnycarney development naturally disrupted existing activities in this semi-rural area, including agricultural, recreational and development uses. This became evident in the public inquiry into the compulsory purchase of the eighty-six acres at Donnycarney, which was held at City Hall in 1937 (*Irish Times*, 17 June 1937). The area covered was bounded – in simplified form – to the north by Killester Lane, to the east by the rears of Killester Gardens and Killester House, and the flank of numbers 12 and 19 Kilbride Road, to the south by the rear of Quarry Cottages and the Clontarf Golf Club, and to the west by the Malahide Road (*Irish Times*, 19 April 1937).

Objections to the purchase give a sense of the various land uses in the broader Donnycarney and Killester areas at this time. Builders Joseph Flynn and William Murphy had previously lodged plans with Dublin Corporation for the construction of 190 houses at 'Kilbride Avenue' (Kilbride Road) and had already completed sixteen houses. In evidence, Murphy stated that he had employed a lot of labour and had sold all the houses thus far completed. The Corporation now planned to take over his partially built road and use it for their scheme. This would deprive him of his profit and force him to look elsewhere for suitable building land (*Irish Times*, 17 June 1937).

A different type of objection was raised by Miss Catherine Byrne and her sister Mrs Leane Cox, who lived at Victoria Park. They were daughters of Tom Byrne, a politician, businessman and successful farmer who had kept prize-winning cattle and bred prize-winning pigs at Donnycarney (*Freeman's Journal*, 28 August 1883). Their holding had originally comprised just over fourteen acres, but the Corporation had already bought more than eight acres. In 1929, the family had spent £500 converting their holding into a poultry farm, managed by Mr Cox (*Evening Herald*, 16 June 1937). The enterprise at Victoria Park was very successful and had become one of the best known such farms supplying poultry throughout the Free State. Indeed, the sisters' white wyandottes won numerous awards in egg-laying competitions throughout the 1930s, with the photo of one prize-winner even featuring in the *Irish Times* (12 December 1935). Mrs Leane P. Cox also won a prize for her shorthorn bull, Desert Nomad, at the Thurles Bull Sale (*Irish Times*, 1 April 1938). In evidence, it was stated that Miss Byrne also grew fruit and vegetables and kept bees. Suitable alternative accommodation for the farm, on which she depended for her livelihood, was not available in the locality. Despite these pleas, Byrne's claim was eventually settled and the land used for housing (*Irish Times*, 18 August 1938). The former site of Victoria Park is now occupied by Clanmoyle Road.

In the drive to build housing, sports grounds in suburban locations around the city were being compulsorily acquired by Dublin Corporation (as at Mount Drummond Avenue and The Thatch, Drumcondra, for example). Similarly, this new scheme demanded the acquisition of several local sporting grounds, including Parnell Park, Killester Park (the sports grounds, not to be confused with the house of the same name) and the adjoining Hollywood Lawn Tennis Club. The proposal would also impact heavily on the Clontarf Golf Club, which had moved from Mount Temple to its current site at Donnycarney House in 1921. Having obtained additional land including McCullagh's Field and Corbett's land (known as the quarry holes) in 1927, the club could boast that it had the only eighteen-hole golf course within a city boundary (Clontarf Golf Club, 2024). However, the new roadway (which became Collins Avenue East) would bisect one of the new greens at Mrs Corbett's land, and the club requested that the road be diverted.

Because of its location relatively close to the city, Killester together with neighbouring Donnycarney had provided sporting facilities from the late nineteenth century. Civil servants from Dublin Castle played rugby and soccer on the playing pitch at St John's Ground, adjoining Victoria Park, which was also used as an athletics venue. Tom Byrne, who himself leased the house and lands of Victoria Park from Dublin Corporation, in turn first leased the grounds for the purpose of Gaelic games in 1901 (Kelleher, 2023: 74). The grounds were renamed as Parnell Park in 1912. In addition to the sporting activities which took place there, a corrugated-iron meeting room which had been transported to the site from Collinstown airfield (now Dublin airport) also served as a community centre and the site became the venue for many concerts and céilithe during the 1920s (Kelleher, 2023: 75). Now with the demand for building land, Dublin Corporation wanted to regain control of the site at Parnell Park

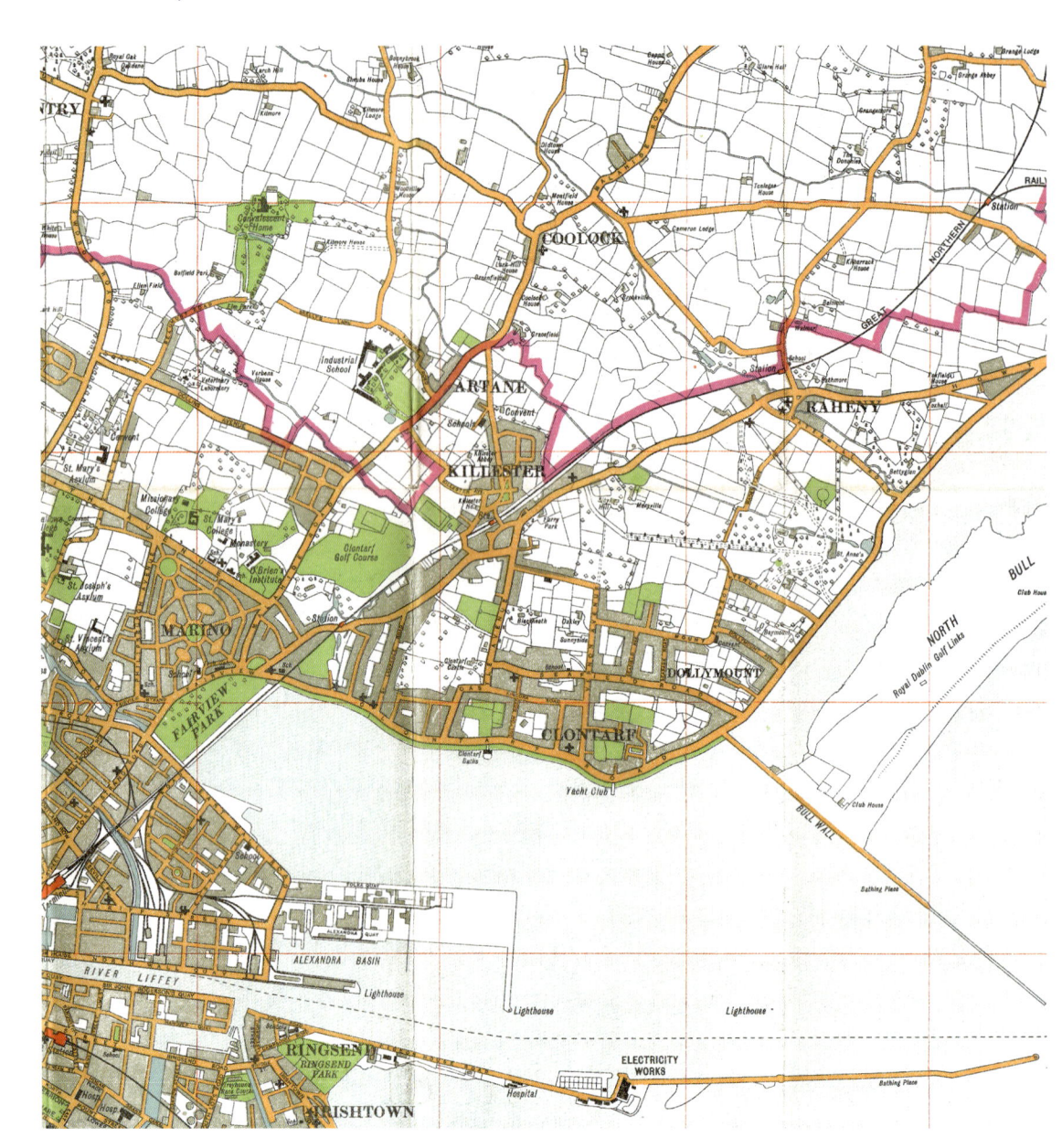

52 Killester and its wider environs. Extract from Ordnance Survey Popular Edition, 1:25,000. The map is dated 1948 but refers to a few years previously. It does not capture the changes being wrought by the extension of Collins Avenue and the development of the 'new' Donnycarney housing scheme. The playing fields at Killester Park and Parnell Park can be seen, while Kilbride Road leads nowhere and the housing development around Furry Park House is not yet in evidence.

The suburbanization of Killester in the twentieth century

for housing purposes. However, the Dublin County Board of the GAA came to an arrangement with the Corporation in relation to its two leases at Killester Park and Parnell Park. The Board withdrew their objection to the acquisition of Killester Park on condition that the Corporation – who were the landlords of the Parnell Park ground – would grant them a new lease of Parnell Park at the existing rent. The negotiations thus saved the Donnycarney playing field at the expense of its Killester neighbour, which was replaced by housing on Collins Avenue East (between Clanhugh and Clanawley roads).

The previous 'old' Donnycarney scheme to the west of the Malahide Road (1929–31) had comprised 421 tenant purchase houses, whereas the newer, larger scheme was for rental. Deputy housing architect Charles P. McNamara explained in 1937 that the new housing area would comprise four-roomed houses at a density of twelve houses per acre, each with a w.c. (toilet) and separate bathroom. There would be one children's playground and three other open spaces. The houses would be served by Killester station (*Irish Press*, 17 June 1937). The houses would be let at 10*s.* per week, whereas the Corporation's previous Donnycarney scheme were being sold to tenant purchasers at higher purchase rents of between 12*s.* 10*d.* and 16*s.* 8*d.* per week. The newer scheme was aimed more directly at former tenement dwellers and, because the houses were rented, the social composition of the new area remained quite different from the older scheme, even as late as the 1970s and 1980s (Brady, 2014). While the houses were small, at 645 sq. ft., they generally had very long back gardens with a smaller garden to the front. The layout involved short terraces on narrow local roads, as there was no expectation that the residents would be car owners.

Part of Dublin Corporation's plans included the extension of Collins Avenue from Malahide Road eastwards to join up with Howth Road. This wide new thoroughfare was in contrast with the existing ancient routeways in the locality, which had mostly derived from farm laneways. Killester Avenue (formerly Killester Lane) had previously been the only connection between the Malahide and Howth Roads, as it meandered from the Malahide Road into the heart of Killester and then out to the Howth Road. Now a far more direct connection between the Malahide Road and Howth Road was made available, reorienting traffic and downgrading the relative importance of Killester Avenue.

The 'new' Donnycarney scheme would bring a significantly increased population and changing social make-up to the area, but its advent was delayed due to the Second World War. Roads and sewers were in place by the autumn of 1942 (Dublin Corporation report 45/1942). However, shortages of materials and workers delayed further development, as did rising costs. Whereas in 1932 the 'all-in' building cost for a cottage had been £430, that had reached £850 by 1945. Nevertheless, work at Donnycarney was underway in 1947 and by 1949 most of the housing in the area now known as 'the Clans' was completed, comprising 890 houses. A shopping parade was

53 Killester and District with the census areas shown. Extract from Ordnance Survey Popular Edition, 1:25,000, 1948 edition. [PC]. The West census area is highlighted in blue and takes in most of Marino as well as the Killester area. The East census area is highlighted in red and reaches as far as St Anne's estate.

provided along the newly extended Collins Avenue East, including the Killester Grand cinema, better known as 'the Killer', which opened in August 1950.

The 1946 census gives a picture of the Killester area just before the new Donnycarney development. Once again, the area was contained within two district electoral divisions, Clontarf West and Clontarf East. Clontarf West included Fairview, Marino and then along the Malahide Road up to Killester Avenue and thus back to the Howth Road with its eastern edge along St Lawrence Road. Clontarf East enclosed most of the Sailors' and Soldiers' Killester scheme but also included most of Clontarf and Dollymount and a great deal of land which was still in agricultural use in the late 1940s.

There had been significant population growth in the eastern DED between 1936 and 1946, but the west had been much more stable. This resulted in a population over both areas in 1946 of 25,494 with 14,737 in the western part. While single-roomed dwellings were an ongoing problem in Dublin city, with 56,912 people still living in one-roomed dwellings, this was much less of an issue in the suburbs. Just 363 people in the two wards were living in one-roomed dwellings, most of them in Clontarf West. Figures for overcrowding were also significantly lower. In line with the pattern across

The suburbanization of Killester in the twentieth century 107

the city, about half of all families consisted of between four and six people. Whereas just under three-quarters of all housing in the city area was rented, this was much lower in the combined areas of Clontarf East and West, at only 37.9% rental. This was due to private housing in Clontarf and Dollymount, but also Dublin Corporation's tenant purchase schemes at Marino and 'old' Donnycarney.

KILLESTER IN THE EARLY 1950S: A SNAPSHOT

A sequence of 1952 aerial photographs show the encroachment of new housing on the surviving older dwellings in the area. Figure 56 shows the modern-day Killester Park housing under construction, with Killester Abbey's old house and barns still intact. The new bungalows, some of which are only partially complete, were designed in 1951 by architect Alfred Jones for the National Housing Society. This was a public utility society that catered for middle-class, white-collar workers, yet the housing, which also included two-storey dwellings, was designed to be of a suitable type to avail of government subsidies and cheap mortgages from Dublin Corporation (under the Small Dwellings Acquisition Acts). A photograph of the laying of the foundation stone of the Society's 106-house scheme at Killester Avenue had featured on the front page of the *Evening Herald* (10 November 1948). The National Housing Society also built residential shops at Killester (*Irish Times*, 14 May 1948). By April 1949 the new three-bedroom houses on Killester Avenue were being sold for £1,650, with a deposit of £150 required to secure a house (*Irish Times*, 1 April 1949). An *Evening Herald* advertisement (15 June 1949) for the National Housing Society's Killester Abbey housing scheme noted that the location was served by the frequent 20A bus, but that garage spaces were also attached to the houses. The Society subsequently built houses at St Anne's Avenue in Raheny.

By 1952, many of the trees once encircling the Killester Demesne had been felled for fuel. However, as Figure 55 illustrates, there was still a rural feel to the locality, with open space and large gardens. Some of the older trees were possibly also nearing the end of their natural lives. A severe storm with gales in September 1950 saw a tree uprooted and crashing through the roofs of two new houses in Killester Avenue, making front-page news (*Irish Press*, 18 September 1950). Fortunately, as the *Irish Independent* reported, the residents escaped uninjured though Mr J. Dalton had a narrow escape, as he was leaving his house carrying his baby when the tree crashed on the roof above his head.

Despite considerable construction activity by the early 1950s, some pockets of land remained untouched. One such site was the former market gardening premises of Killester Gardens and its neighbour, Killester House (not to be confused with the older Killester Demesne house which was by now long demolished). The second Killester

54 Aerial view of new Collins Avenue housing schemes and Killester in 1952. The image looks towards the port and the line of the Bull Wall and the South Wall. [XAW044955. HES]

The suburbanization of Killester in the twentieth century 109

55 Aerial view of new Collins Avenue housing schemes and Killester in 1952, north-west. This image offers a view of the same area as Figure 54 but from a different angle, towards the north-west. The line of the Malahide Road divides the image at the middle, while the line of the railway is clearly seen towards the bottom of the image. [XAW044960. HES]

56 Aerial view of Killester Park under construction in 1952, with the house known as Killester Abbey in the foreground. [extract from image XAW044955. HES]

57 The former water tower at the ex-servicemen's estate is visible near the treelined 'nuns' walk' in this extract. There is a suggestion of unevenness in the ground between the trees and the Legion Hall, perhaps indicating the remains of the former demesne house below the surface. [extract from image XAW044960. HES]

The suburbanization of Killester in the twentieth century

58 This enlargement shows the railway line at the top and Killester Avenue in the foreground. The house surrounded by walled grounds is the former Killester House (present-day site of The Bramblings). Recently completed two-storey semi-detached houses line Killester Avenue. With their distinctive layout of three upstairs windows per house these contrast with the new Dublin Corporation housing on Clanawley Road, which begins with the distinctive angled house treatment on the corners. [extract from image XAW044955. HES].

House had been occupied by the Cairnes family from the turn of the twentieth century. The aerial views from the early 1950s (see Figures 54 to 58) reflect much of how the house was described when the freehold was put up for sale in 1937. The 'exceptionally charming' house included grounds of 2 acres and 7 perches, including a tennis court and beautiful flower and rock garden. The adjoining property (unnamed but presumably Killester Gardens) was also available, if desired. This consisted of a two-acre plot with a walled orchard, greenhouse, vinery and dwelling (*Irish Times*, 17 February 1937). A further advertisement noted that a high wall in good condition surrounded the entire property (*Irish Times*, 9 March 1937). This wall is still in evidence in the aerial view and portions of it survive to the present day as garden walls. At this time the house appears to have been acquired by the Murray family, who hosted a wedding party there in 1944, which was described in the newspaper society pages.

KILLESTER HOUSE, KILLESTER.
DELIGHTFUL FREEHOLD RESIDENCE, WITH 2 ACRES 7 PERCHES.

This exceptionally charming non-basement House, which a generous expenditure, controlled by refined taste, has made one of the most attractive homes in the northern suburbs of Dublin, stands in slightly over 2 acres of ground, including a tennis court, a rarely beautiful flower and rock garden, peach house, greenhouse and carpenter's workshop, and pleasure grounds.

The House contains lounge hall, with fireplace and cloakroom, drawingroom, study, liningroom with conservatory off, verandah, seven bedrooms, dressingroom, bathroom with fixed basin and heated towel rail, 2 servants' rooms, tiled kitchen with Ideal boiler, larder, pantry, with hot and cold water, etc.; heated linen cupboard. The offices include chauffeur's living room and bathroom, garage with sliding gates, tool shed, potting shed, etc. Electric light and gas; corridor radiator.

The entire held for ever, free of rent, with possession to purchaser. The adjoining Premises, consisting of a walled orchard, greenhouse vinery—in all, about two acres, with a Dwelling—all held in fee simple, will be included in Sale if purchaser desires. A high wall in good condition surrounds the entire property.

W. S. BARRETT, Solicitor, 15 Sth. Frederick street,

JAMES H. NORTH & CO., AUCTIONEERS, 110 GRAFTON ST.

59 The sale of Killester House in 1937. *Irish Times*, 2 March 1937: 16.

When Killester House was sold once again in 1945, the changing character of the surrounding area was reflected in the particulars. The property was in two lots, one of which included the house and grounds, while the second lot comprised the former Kirby market gardening holding (Killester Gardens). This latter two acres with orchard, vinery, tomato houses, large yard and three sheds was 'adaptable for market gardening'. However, the newspaper advertisement also noted that 'these lands have frontages of about 630 feet on Killester Avenue and are ripe for building' (*Irish Times*, 10 March 1945). By the time that the property changed hands again in 1955, that frontage had been availed of for housing purposes (as seen in Figure 60). The owner John J. Farrell, a former lord mayor of Dublin and a pioneer of the film business in Ireland, had died at Killester House in June 1954 (*Irish Times*, 3 June 1954). Part of the executor's sale included ground rents of £124 per annum arising out of numbers 16, 18 and 72–98 Killester Avenue (even numbers). Clearly these 16 subleases related to the new houses which had been built along the edges of the property (*Irish Times*, 9 November 1957). Meanwhile the house with approximately three acres of land was being offered for sale with a note that it would suit a private residence or nursing home (*Irish Times*, 15 February 1955), while 'it would be possible to develop the lands

The suburbanization of Killester in the twentieth century

60 This extract shows the surviving Killester House which is at right angles to the roadway behind a high wall. There is a glasshouse along one wall of the site, perhaps a remnant of the former market gardening activity at Killester Gardens. The latter house appears to have been demolished at this stage. [extract from image XAW044960. HES]

61 The nun's walk in 1943. [ITA-Coolock-04. DCLA]

without affecting the quiet and seclusion of the residence'. In the event, the house survived until the 1970s. It became a private convalescent home run by Norah Caffrey, whose husband Thomas (mentioned above) built many houses in the Howth Road and Clontarf area.

SHOPS AND SERVICES FOR A GROWING POPULATION

One feature of the rapid growth of population in the locality was an increase in demand for groceries and other goods and services. Some shops had been constructed along the Howth Road in the 1920s, although as late as 1940 the shopping parade was not fully complete. Number 169 Howth Road (which by 1971 had become H. Williams supermarket) was still listed in the street directory as building ground, but other premises were occupied by a pharmacy, Kennedy's bakery, a newsagent, pork butcher, ophthalmic optician, victualler, hardware stores and 'Killester Cash Stores' and 'Killester Stores' (Thom's Directory, 1940). By the 1940s 'the tin shops' adjacent to La Vista Avenue were also plying their wares. The exterior of Jimmy Byrne's dairy and grocery shop featured cigarette advertisements, while children who grew up in the area recollect the comics and ice creams that they purchased there. Dublin Corporation also built six residential shops at Killester Avenue, which were 'a unique opportunity to procure a business in a densely populated and growing area' according to the sale advertisement in 1947 (*Irish Times*, 16 September 1947). The location of the shops was described as being 'in the centre' of the Corporation housing area and adjoining several other housing schemes in progress and about to commence. A publican's license was granted to Brian O'Shea of 145/7 Killester Avenue in 1952. The character of Killester Avenue had irrevocably changed.

A clearance sale of the dairy herd at Rosedene, Killester Avenue, was advertised in the *Irish Times* on 12 January 1948. Although the dairying activities that had previously taken place in Killester had given way to housing by the early 1950s, the growing population resulted in a rising demand for milk and other dairy products. As a result, Merville dairies established a depot at Killester from which milk was delivered to the surrounding new housing areas using horse-drawn vehicles. When all thirty horses in the Killester depot were affected by a flu virus, the *Irish Times* reported that motor transport had to be used on their rounds, but that this was less suitable and more expensive than horse transportation. A Merville spokesman observed that horses were more suitable for the suburbs, as well as being considerably more economical. An added benefit was that 'they follow the deliverymen from house to house on their rounds, something the vans cannot do!' (*Irish Times*, 30 March 1965: 14). The Merville/Premier dairy depot at Killester was home to one of the last horse-drawn milk delivery cars in Dublin, which finally went out of service in 1978 when its operator retired. The dairy depot at Killester, by now part of Premier dairies, survived until the mid-1980s. In May 1984 the Premier-Hughes group announced that it would cut 600 jobs and close two of its five Dublin depots, at Killester and Kimmage. By summer 1985 the depot had closed, although the site would remain vacant for some time, before being replaced by the Killester Court complex in 1996. The twenty-five-house scheme and community centre (Sonas Hall) was backed by Women's Aid and specifically provided housing for victims of domestic violence (*Irish Independent*, 12 November 1996).

With the many young people growing up in new housing in Killester and its surroundings, additional medical and educational facilities were needed. Tenders for the erection of a dispensary (now health centre) at Killester Avenue were sought by the Dublin Board of Assistance in 1950 (*Irish Times*, 14 November 1950). In 1954 work began on Killester Vocational School, which was completed in 1956. Designed by architects Hooper and Mayne, it is considered to be typical of mid-century Irish modernist architecture (Rowley, 2018: 443). The intention was to provide vocational education locally, rather than requiring pupils to travel to the city centre. Similar schools were also opened at Emmett Road and Crumlin Road during this period. The 'tech' became a college of further education. Since 2023 the building has housed the Killester Raheny Clontarf Educate Together National School (KRC ETNS). Grounds behind the convent on St Brigid's Road were used for St Mary's secondary school for girls, which opened in 1967.

Given the scale of housing development in Killester after the Second World War, the Roman Catholic St Brigid's church, which had opened in the late 1920s, was too small for the growing population. Architectural practice Robinson, Keefe and Devane, successor to J.J. Robinson who had designed the original structure, planned and supervised the extension project. The work was carried out by builder John Lambe of North Strand from 1950 to 1952. Already in mid-1947 work began on the erection of a chapel of ease for Marino – 'The Tin Chapel' – which was consecrated in March 1948 at Donnycarney. A new Catholic parish of Donnycarney was constituted in 1952 from the parish of Marino, but a permanent church was not completed until 1969. New schools for boys and girls were also built along Collins Avenue East.

When the Killester Cinema was opened by Jimmy O'Dea, deputizing for the lord mayor of Dublin, in 1950, few could have imagined that its life as a picture-house would span only two decades. It was one of a new generation of large-scale cinemas, like the Gala in Ballyfermot and the Cabra Grand, which opened in the late 1940s and early 1950s to serve a growing suburban population. Built by contractor D.A. McCambridge to designs by architects Munden & Purcell, the 'Killer' had a seating capacity of 1,250. Going to 'the pictures' was still a hugely popular form of entertainment in the 1950s, with programmes changing regularly. While continuous programmes ran from 3 p.m. daily, there were two separate showings on Sundays and

62 The 'Killer' shortly before its demolition. [DCC].

booking was advisable for Sunday night. Adult tickets ranged in price from 1s. 3d. to 2s. 2d. depending on time and location within the auditorium – with balcony seats at an evening showing being the most expensive. For children, a seat in the stalls at a matinée showing cost just 4d. It was frequently remarked that the entertainment within the auditorium often equalled anything on the screen! A dramatic burglary in 1968 saw thieves getting away with 20,000 cigarettes from the Killester Cinema, having scaled two twenty-foot-high walls, removed glass from a skylight, dropped eighteen feet onto the cinema stage and then bored a circle around the lock with a bit-and-brace in order to access the foyer where the goods were stored (*Evening Herald*, 20 January 1968).

By September 1970 the doors of the Killester Cinema, along with those of the Rialto, also owned by the Green Cinema group, had closed 'as a result of spiralling wage bills' (*Irish Press*, 11 September 1970). Whereas the Casino in Finglas, also part-owned by the Green Group, quickly found a new life as a supermarket, the Killester cinema had a rockier path. A 1971 campaign by the local Donnycarney community to have it converted into a community centre failed, and the cinema was put up for sale. This process took several years, however. It reopened briefly for bingo in 1973, and was occasionally used for musical acts, including a performance by 'top international act' White Plains on 21 May 1973 (with admission at 50p, 60p and 75p), but in May 1974 was described as being 'long unused' (*Evening Herald*, 21 May 1973; 16 May 1974). Late in 1977 permission for a change of use was sought, and in 1978 the former cinema gained a new lease of life as a camping showrooms. At 19,000 square feet the O'Meara's showrooms was 'one of the biggest shops of its type in Europe' when it was officially opened by the minister for tourism and transport, Pádraig Faulkner (*Irish Independent*, 19 May 1978). Part of the former cinema balcony was set aside as a separate hall with

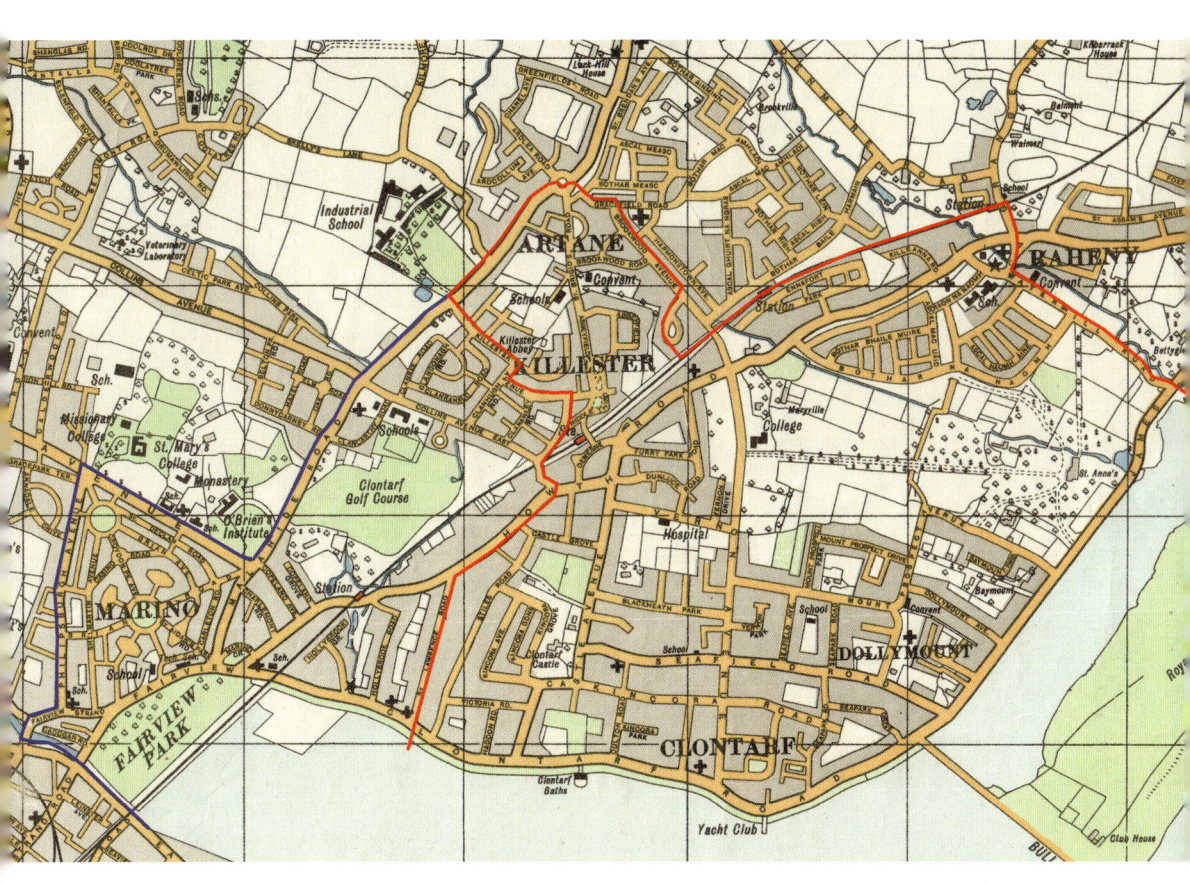

The suburbanization of Killester in the twentieth century

63 The Ordnance Survey Popular Edition, published in 1959, shows the dramatic changes brought about by Dublin Corporation's new Donnycarney housing scheme and extension of Collins Avenue East, including its shops and cinema. The private Craigford estate and the National Housing Association development on Killester Avenue and Killester Park is also visible, although the latter road is incomplete. Relatively few pockets of unbuilt ground remain, such as around Killester House (later the Bramblings) and Venetian Hall. To the north-east, Brookwood Avenue and the Harmonstown area has been completed. The 'tree-lined walk' around the servicemen's housing, connecting with the convent grounds, can also be seen. The census areas used for the 1961 census are outlined in colour – blue for Clontarf West and red for Clontarf East.

projection facilities. This 88-seater venue was made available free of charge to community groups (*Evening Herald*, 22 May 1978). On Friday nights in March 1979 a free holiday film show was advertised. However, by 1981 an application had been made for a change of use of the former balcony to a snooker club and coffee bar. Given that the popularity of television was a key reason for the demise of suburban cinemas, it is somewhat ironic that by 1983 the Access Video Club was operating from the venue, offering £25 life membership for access to an extensive collection of VHS videos for home viewing (*Irish Independent*, 14 January 1983).

KILLESTER IN THE 1961 AND 1971 CENSUSES

The development of 'new' Donnycarney and the substantial private housing schemes in Killester during the late 1940s and 1950s are reflected in the findings of the 1961 census. Between the 1946 and 1961 censuses there were some small changes to the boundaries of the wards to bring them more in line with streets. This meant that Clontarf East went as far as the recently completed Artane roundabout before heading south via Gracefield Road and Brookwood Avenue. The population had seen significant growth to 40,132, most of it occurring in the early years of the 1950s. There were some small and subtle changes to family structure, mainly a result of changes in the life cycle of the area. Households of four, five and six people still accounted for half of all households in Clontarf East and forty per cent in Clontarf West. The relative importance of one- and two-person households increased in both areas. In the east area it changed from around eight per cent to twelve per cent but there was much more dramatic change in the west where the relative importance almost doubled from fourteen per cent to twenty-six per cent. Unfortunately, it is not possible to determine the causes of this change – was it due to new arrivals, young couples who had not yet begun their families, or does it reflect empty nests as grown children move out of the family home? Overall, overcrowding in the city had declined to 14 per cent of the population in 1961, and the rates in the Killester area were even lower. Very few people in Clontarf East (1.6%) were overcrowded, whereas just under nine per cent of the population in Clontarf West was overcrowded, largely a result of the increase in social housing where often large families occupied standard-sized four-roomed dwellings.

The 1961 census provided additional information on housing tenure. Dublin was still a city of renters where 62.7 per cent of dwellings were rented, 34 per cent from the private sector. The two areas in Clontarf had markedly different profiles. In Clontarf East, most dwellings (79.3%) were owner occupied and hardly any were rented from Dublin Corporation. In contrast, for Clontarf West, owner occupiers accounted for 45.7% of dwellings and tenant purchasers (who were buying their houses from Dublin Corporation) accounted for a further 15.5%. The tenants of the 'new' Donnycarney scheme accounted for the 21.8% of rental from Dublin Corporation, with 'other private rental' accounting for the remaining 13.2%.

In Dublin city as a whole, just over half of all dwellings had been built since the foundation of the State with one quarter built in the years since 1948. Most of the housing in Clontarf East and West was even newer. Over 80 per cent of dwellings had been built since 1919 but almost half of those in the east were post 1948 and nearly 40 per cent of those in the west. It is unsurprising, therefore, that almost all dwellings had a fixed bath and an indoor flush toilet. This contrasted with the city as a whole where, even as late as 1961, almost one in five dwellings did not have an indoor toilet and only seventy per cent had a fixed bath.

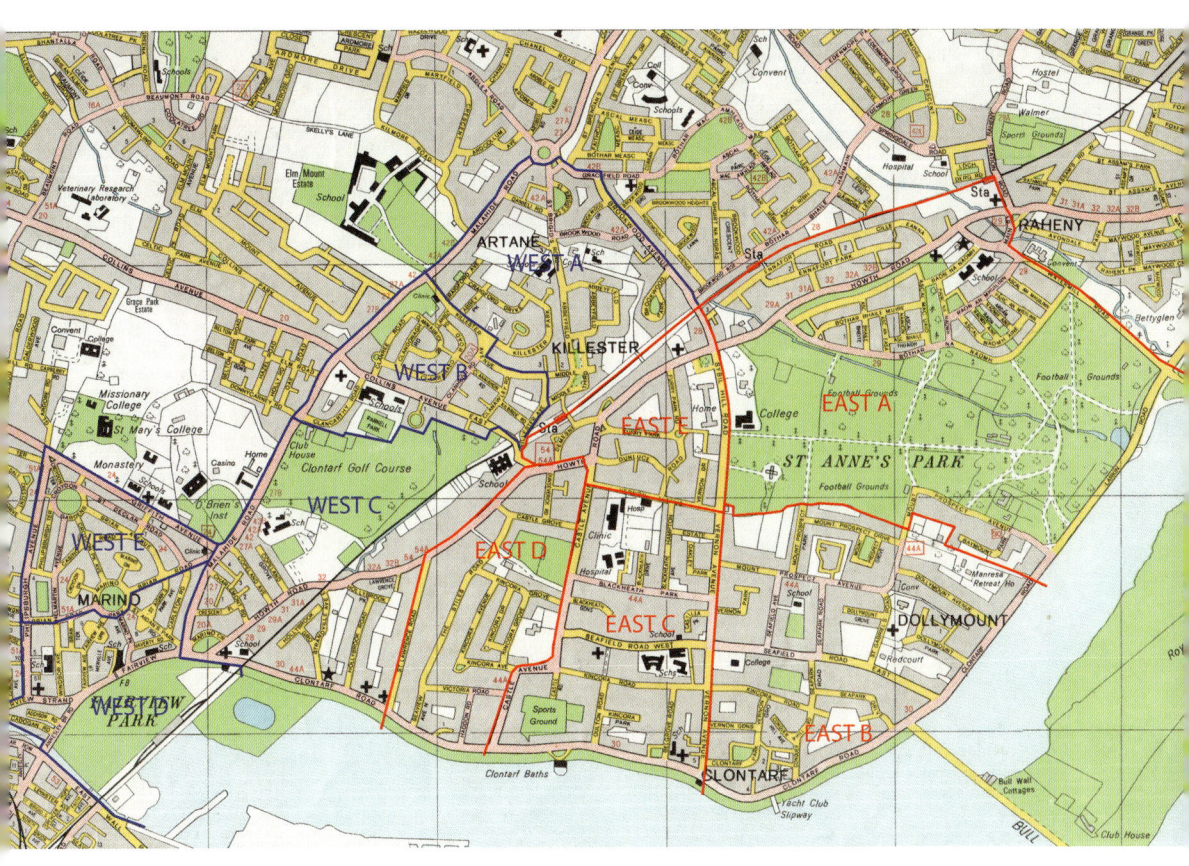

The suburbanization of Killester in the twentieth century

64 Killester and environs in the early 1970s as show in this extract from the Ordnance Survey Popular Edition 9, published in 1974. [PC]. The revised census areas used in the 1971 census are overlain in colour. There were now five areas in each of what had been Clontarf West and Clontarf East.

For the 1971 census, a new set of boundaries was introduced to give greater granularity to the census data. These smaller statistical areas have proved very valuable in understanding the city. Clontarf West and East were each divided into five sub-units, designated A, B, C, D, E (see the various Popular Edition maps). Three sample areas are considered here, to give a sense of the varied nature of Killester (and environs) by 1971. The area of Clontarf West B was largely occupied by the 'new' Donnycarney scheme, but also included Kilbride Road and Killester Avenue (see map). By contrast, Clontarf East E included Demesne, part of the Howth Road between the junction with Castle Avenue and the junction with Sybil Hill Road, and the Furry Park Road estate, Dunluce Road and Vernon Avenue. Finally, Clontarf West A included Middle Third, Abbeyfield, Killester Park, Craigford, St Brigid's Road, extending as far north as the Artane Roundabout and east to Brookwood Avenue.

The table below captures some of the variation in character between these areas, as represented in the 1971 census. While Clontarf West A and B were of comparable size, their housing tenure contrasted markedly. Just over three-quarters of housing in Clontarf West B was being rented from the local authority, reflecting the scale of the 'new' Donnycarney scheme, with a low ownership rate of just over 6 per cent. Its residents were most likely to be in the skilled manual (24.79%), semi-skilled manual (20.33%) or unskilled manual (18.18%) occupational categories. By contrast, over four-fifths of all homes in Clontarf West A were owner occupied. The residents were largely in the census occupational categories of skilled manual (30.15%), intermediate non-manual (27.32%), other non-manual (12.56%) or employers (10.43%). The less populated Clontarf East E had the highest ownership rate (at 87.26%) but also had the highest rate of private rental housing than either of the other two wards (9.46%). Here the greatest proportion of residents fell into the occupational category of intermediate non-manual (37.78%), followed by employees (16.9%) and skilled manual (14.01%). Of the three selected areas, Clontarf East E had the greatest proportion of higher professionals (7.01%). This contrasted with 2.43% higher professionals in Clontarf West A and just 0.31% in Clontarf West B.

Selected census indicators for small areas in 1971.

1971 Census	Clontarf West B	Clontarf West A	Clontarf East E
Total population	4510	4315	1784
Housing units	966	965	518
Types of tenure			
Local Authority rental	76.7%	4.6%	0%
Private rental	1.0%	5.3%	9.5%
Being purchased from Local Authority	15.7%	2.6%	0.2%
Owner occupied (including mortgaged)	6.3%	84.4%	87.3%
Cars per household	0.26	0.67	0.69
Per cent non-Roman Catholic	3.6%	8.0%	12.2%
Per cent Irish speakers	13.3%	35.6%	33.7%

The mature Killester area now boasted a number of shopping locations. The Howth Road cluster included several supermarkets (H. Williams, Power and Lipton), the Hibernian Bank, public house (The Beachcomber), chemist, two victuallers, pharmacy, dental surgeon, tobacconist, IMCO cleaners, a coin laundry and separate hairdressers for men and women. At Collins Avenue East the shopping parade at the cinema included hairdresser, victualler, chemist, turf accountant, dry cleaners, a grocery, and two confectioner-newsagents. Finally, the small cluster of shops at

Killester Avenue included Rowley's grocery and sub-post office, the Ramble Inn public house, Puzzuoli's fish café, Terminus Stores grocery, Lynam's victuallers and hairdressing salon, Lowe's grocery and Quill's chemist (Thom's directory, 1971).

FINAL INFILL DEVELOPMENT IN KILLESTER AND THE 1991 CENSUS

A third and final significant phase of development in the 1970s and 1980s saw infill development of the last remaining large house sites in the Killester area, including Venetian Hall and Killester House. The type of development was also changing. Apartments were increasingly in vogue, while more compact 'townhouses' were being built to make the most of relatively small sites. Outline planning permission for the 4.5-acre Venetian Hall site at 276 Howth Road had been sought by T. Walsh as early as 1971, but it was not until November 1975 that the final go-ahead for development was granted by the minister for local government (*Irish Times*, 10 September 1971; 7 November 1975). Whereas the initial planning permission in 1971 was for a single three-storey block, by the time of construction this was one of Dublin's biggest flat-block developments, planned to offer 180 units across seven blocks varying in height from two to five storeys. The apartments were built to high standards, although the features noted in a display advertisement from Sheelin Homes in 1981 may seem somewhat quaint today. The flats due for completion in January 1982 would cost from £37,250 with no stamp duty and boasted private balcony, fitted carpets, net curtains, gas fired central heating, double glazing and a brickwork open fireplace.

Venetian Hall caused quite a stir in the property press. By the time that the sixth phase of the development was launched in 1984, it was being described as 'the first real American-style condominium north of the Liffey' (*Irish Times*, 3 February 1984). A new bridge over the railway line had been built to access the mix of apartments and duplexes. Features included a rooftop conservatory garden with extensive views, remote-control garages, floodlit hard tennis courts and putting greens. These modern conveniences were considered 'cutting edge' at the time, much as the original 'neat commodious house' known as Venetian Hall had been when it was first advertised in the *Dublin Courier* in 1760, just a year after its construction.

Killester House was used as a convalescent home and was finally demolished in the 1970s. Initially the owner, T. Caffrey, sought planning permission for an apartment development comprising six three-storey blocks with 96 units in total (*Evening Herald*, 4 August 1973). Subsequent modified applications were for Carroll System Buildings Limited, with permission granted for eight three-storey blocks by October 1974. However, this development did not proceed and instead Carroll System Buildings sought permission for a two-storey shopping centre on the site (*Irish Independent*, 18 September 1975). The character of present-day Killester Avenue would be quite different had this gone ahead. Instead, the site was developed into townhouses, and by the end of 1978, new three-bedroom red-brick houses built by Tower Homes were being advertised at 'The Brambles', Killester (*Irish Independent*, 9 December 1978), in an 'exclusive cul-de-sac'.

The 1990s saw the development of the last substantial tract of unbuilt land in Killester, when the St Brigid's estate off St Brigid's Road was completed. These changes are not reflected in the 1991 census data presented below, which gives a sense of the changes in population, tenure and housing type which had taken place in the twenty years since 1971. Of note here is the significant increase in home ownership in Clontarf West B, as former Dublin Corporation tenants availed of 'right to buy' schemes. Whereas in 1971 over three-quarters of houses in the area were being rented from the local authority, this had fallen to just 17.65% in 1991. Across Dublin county borough (the area controlled by Dublin Corporation), the population had fallen by 4.8% between 1986 and 1991. This was a period of high unemployment and emigration. The total population in all three areas had also fallen, partly due to emigration but also likely due to the changing demographic structure, as the large young families of the early 1970s had given way to empty nesters and incomers with smaller average family sizes. This drop in population is evident even in Clontarf West A, which had seen an increase in the overall number of housing units (from 965 to 1,175). Most of these additional housing units were at Venetian Hall. One outcome of the newer infill development which included apartments was that there was a decrease in the proportion of conventional houses at Clontarf West A (88.41%, compared with 96.46% at Clontarf West B and 95.46% at Clontarf East E).

Selected census indicators for small areas in 1991.

	Clontarf West B	Clontarf West A	Clontarf East E
Population 1986	3160	3442	1655
Population 1991	2934	3290	1579
Per cent change 1986–91	−7.2%	−4.4%	−4.6%
Housing Units	986	1175	523
Built 1971–81	38	70	20
Built since 1981	32	182	16
Types of tenure*			
Local Authority rental	17.7%	4.1%	0%
Private rental	1.9%	6.4%	7.7%
Being purchased from Local Authority	29.0%	0.9%	0%
Owner occupied (including mortgaged)	50.4%	87.7%	90.4%
Per cent conventional housing	96.5%	88.4%	95.5%
Per cent of labour force at work	76.2%	91.7%	95.4%

*note that the percentages do not sum to 100% as the category 'occupied free of rent' and 'not stated' are not included in this table.

The suburbanization of Killester in the twentieth century

65 Holy Faith convent in 1943. [ITA-Coolock-05. DCLA]

Although the final infill phase outlined above replaced the last of the historic houses and farms in Killester, incremental infill at a much smaller scale has continued in more recent years. This has primarily involved the construction of single houses in the gardens of houses with a large garden site. Long back gardens have also been developed with rear extensions, and former single-storey garages have been converted into living space, sometimes including an upper storey. Since 2017, however, a different type of development has been undertaken in Killester, involving redevelopment and densification of two substantial sites, at the former St Brigid's convent and at the former Killester Cinema. In both cases the development was disputed and An Bord Pleanála was involved in adjudicating the outcome.

In 2017, coincidentally the fiftieth anniversary of the secondary school, the Holy Faith order sold their former convent on 2.2 acres. Following an initial rejection by Dublin City Council, planning permission was granted by An Bord Pleanála for a ninety-unit housing development at the site. An additional storey was added to the convent, which was refurbished and converted into apartments, while a further two blocks of apartments and four three-bedroom houses were completed on the grounds. This controversial development is the most significant single transformation of part of the former Killester Demesne since the erection of the ex-servicemen's houses almost a century before. Meanwhile on Collins Avenue East, the long-awaited redevelopment of the former cinema site is underway at the time of writing (July 2024). This new

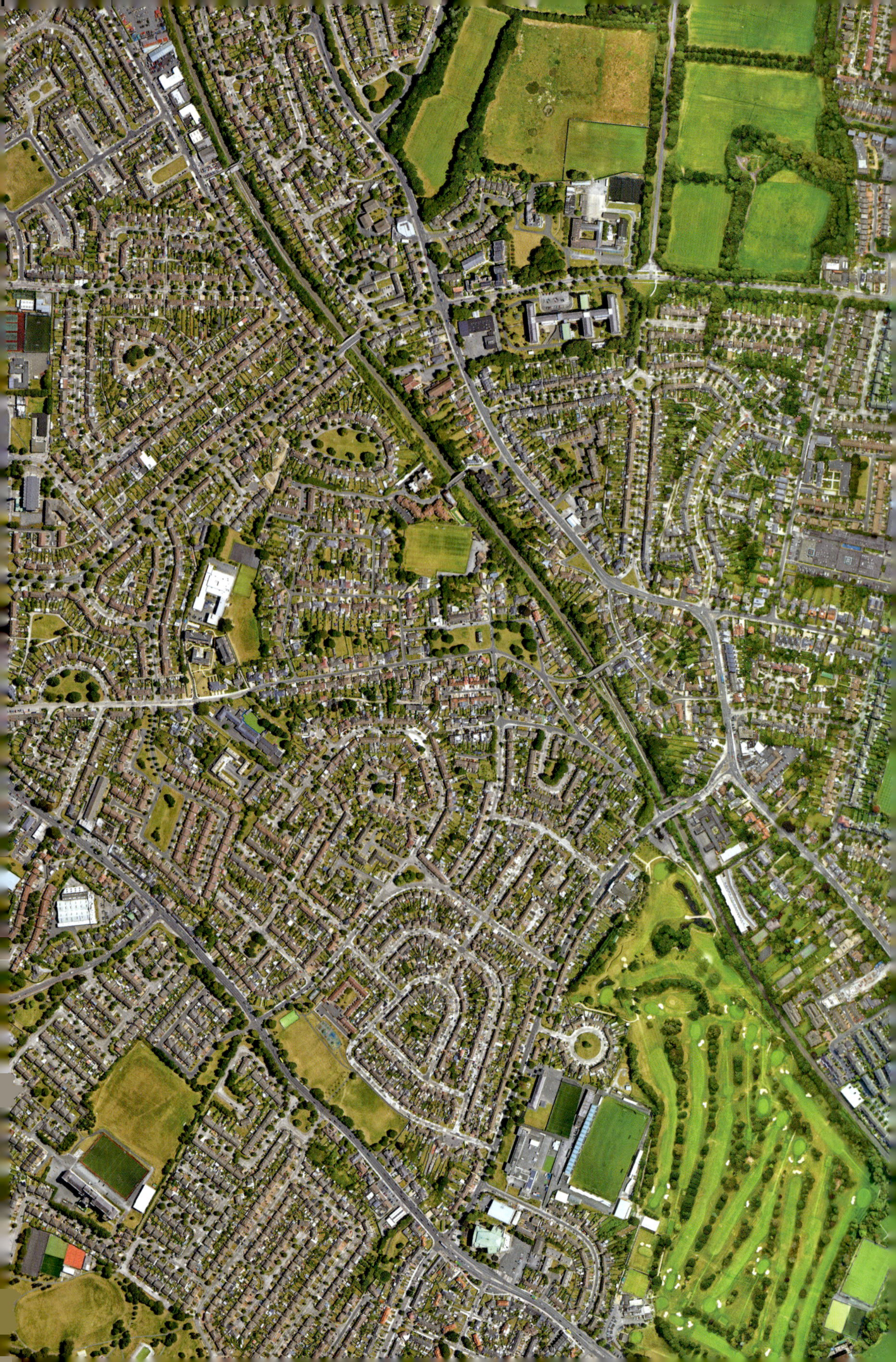

apartment complex has been named Easra Court and will consist of thirty-three one-bedroom and thirty-four two-bed apartments over seven storeys.

Both of these recent developments, 'Brookwood Court' (86 apartments, 4 houses) and 'Easra Court' (67 apartments), have been undertaken by the same developer, MKN Property Group, and all are 'build to rent' units. This is a relatively new concept in Ireland, which involves purpose-built residential rental accommodation and associated amenity space that is professionally owned and managed by an institutional landlord. While it was still under construction, it was reported that the Brookwood Court scheme had been acquired by a real estate fund managed by Deutsche Bank subsidiary DWS, as part of a portfolio of 317 residential units completed by the MKN Property Group. The scheme had a gross estimated rental value of €2.27 million per annum (*Irish Times*, 29 July 2020). Unlike the older housing in Killester, both new developments are intended for rental accommodation in the long term, without any future opportunity for residents to purchase their homes.

REFLECTION

The incremental suburbanization of Killester in the century since the completion of the scheme for ex-servicemen is not unique to this area. Similar processes have operated in locations around cities in Ireland and further afield, as the relationship between rural, urban and suburban areas gradually shifts and evolves over time. However, each suburban area retains a unique character which reflects the multiple phases of development and the many generations of people who have lived there. Writing in 1929, when a former walled garden at Killester Demesne was due to be developed for housing (as The Orchard), a newspaper columnist opined that the existing tenants should be encouraged and educated to use these old gardens to produce vegetables or even flowers, rather than building houses on the site: 'Some day we shall begin to realize that open spaces are quite as necessary as bricks and mortar. Killester, in this respect, is the nearest approach that I have seen to the ideal. Why not preserve it?' (*Irish Times*, 10 July 1929). While Killester is now densely developed for the most part, the open spaces in the area are now valued to a greater degree than ever before. The recently established Killester Blooms Community Garden, occupying a narrow strip of land between Abbeyfield and St Brigid's girls' national school, has become a well-loved and award-winning space which symbolizes the community spirit and continued evolution of Killester.

66 Killester and environs in a recent image captured by Google Earth.

Notes

CHAPTER 1. *The evolution of Killester from earliest times*

1. https://www.logainm.ie/en/1415880.
2. https://www.duchas.ie/en/cbes/4498837/4385812.
3. Confirmation by Cardinal Vivian of a grant to William Brun by the Canons of Holy Trinity, Dublin, of the land called 'Killastre' (Killester, Co. Dublin), 1178 April 14; Fee farm grant from the Canons of Holy Trinity, Dublin, to William Brun, of the land called Killastre, 1178–1180; Confirmation by Laurence (O'Toole) Archbishop of Dublin of a fee farm grant from the Canons of Holy Trinity, Dublin, to William Brun of the land of Quillestre, 1178–1180.
4. https://www.dib.ie/biography/st-lawrence-robert-a8222
5. https://catalogue.nli.ie/Record/vtls000741758
6. https://civilrecords.irishgenealogy.ie/churchrecords/images/deaths_returns/deaths_1903/05672/4597502.pdf
7. https://civilrecords.irishgenealogy.ie/churchrecords/images/deaths_returns/deaths_1909/05445/4523263.pdf
8. https://civilrecords.irishgenealogy.ie/churchrecords/images/deaths_returns/deaths_1887/06198/4772362.pdf
9. http://www.census.nationalarchives.ie/pages/1901/Dublin/Clontarf_East/Killester_Demesne/1269700/
10. http://www.census.nationalarchives.ie/pages/1901/Dublin/Clontarf_East/Killester_Demesne/1269701/
11. http://www.census.nationalarchives.ie/pages/1911/Dublin/Drumcondra_Rural/Killester_Demesne_Part_of_Rural_/7832/
12. http://www.census.nationalarchives.ie/pages/1911/Dublin/Drumcondra_Rural/Killester_Demesne_Part_of_Rural_/7836/

CHAPTER 2. *The Killester Garden Village*

13. It was the practice of Dublin Corporation to number committee reports and bind them into annual volumes. A reference to 'Report xx/yyyy' is to a report numbered xx for year yyyy.
14. There is considerable variation in the use of the apostrophe in the title of the Sailors' and Soldiers' Land Trust in both the legislation and the Trust's documentation.
15. The archive of the Sailors' and Soldiers' Land Trust is held in the UK National Archives at Kew, Richmond, London. However, it contains little on the activities of the Local Government Board prior to the Trust's establishment. References to Trust documents in the text above are to the file numbers in Kew.
16. The practice of giving bus services individual names harked back to the beginning of the omnibus service when it comprised horse-drawn coaches. It became feasible to assign numbers to routes only when the three main horse tramway operators, the Dublin Tramways Company, the North Dublin Street Tramways Company, and the Dublin Central Tramways Company merged in 1881 to form the Dublin United Tramways Company (DUTC). Retaining the practice of 'catchy' names allowed much smaller scale operators to maintain a distinctive identity. The name of the 'Contemptible' bus company was not an insult but rather a commemoration of the British Expeditionary Force which served with distinction in Flanders early in the First World War, outnumbered and with little equipment. They became known as the 'Old Contemptibles', a reference it seems to an order of the day from the Kaiser which referred to them as a 'contemptible little army'.

Bibliography

Aalen, F.H.A. (1988). Homes for Irish heroes: housing under the Irish Land (provision for soldiers and sailors) Act 1919 and the Irish Sailors' and Soldiers' Land Trust, *Town Planning Review*, 59(3), 305–23.

Archaeology Plan Heritage Solutions (2023). Archaeological assessment of former demesne lands, Killester, Dublin 5. Unpublished report for Dublin City Council Heritage Office.

Archdall, Mervyn (1786). *Monasticon Hibernicum, or, An history of the abbies, priories, and other religious houses in Ireland. Interspersed with memoirs of their several founders and benefactors, and of their abbots and other superiors, to the time of their final suppression.* Dublin: Luke White. Vol II. Edited by Patrick F. Moran and reprinted 1876.

Archer, Joseph (1801). *Statistical survey of the County Dublin.* Dublin: The Dublin Society. Available at https://www.askaboutireland.ie/reading-room/digital-book-collection/digital-books-by-county/fingal/dublin-statistical-survey/

Armstrong, E.C.R. (1915). Descriptions of some Irish seals, *The Journal of the Royal Society of Antiquaries of Ireland*, 5(2), 143–8. http://www.jstor.org/stable/25514401

Ball, Francis Elrington (1917). *Howth and its owners.* Dublin: The Royal Society of Antiquaries of Ireland.

Bowen, B. (1963). South-east Fingal. *Dublin Historical Record*, 18(3), 66–79. http://www.jstor.org/stable/30083947

Brady, J. (2014). *Dublin, 1930–1950, the emergence of the modern city.* Dublin: Four Courts Press.

Brady, J. (2016). *Dublin, 1950–1970: houses, flats and high rise.* Dublin: Four Courts Press.

Brady, J. and McManus, R. (2021). *Building healthy homes – Dublin Corporation's first housing schemes 1880–1925.* Dublin: Dublin City Council and Four Courts Press.

Brady, J. and Lynch, P. (2009). The Irish Sailors' and Soldiers' Land Trust and its Killester nemesis, *Irish Geography*, 42(3), 261–92.

Brewer, J.N. (1825). *Beauties of Ireland.* Vol. I. London: Sherwood Jones & Co.

Burke, J. (1846). *A general and heraldic dictionary of the peerages of England, Ireland, and Scotland.* 3rd edition. London: Henry Colburn.

Cannon, Séamus (1985). *Furry Park House, a short history.* Dublin: Furry Park Action Group.

Carolan, Noel (2018). A local war-time food supply initiative: the Clontarf and Marino allotments of 1917, *Dublin Historical Record*, 71(2), 223–33.

Casey, Christine (2005). *Dublin.* USA: Yale University Press.

Clontarf Golf Club (2024). History of Clontarf Golf Club, https://www.clontarfgolfclub.ie/about-us/club-history/, accessed 24 June 2024.

Colgan, N. (1895). The orchids of County Dublin, *The Irish Naturalist*, 4(8), 193–8. http://www.jstor.org/stable/25520847

Cosgrave, Dillon (1909). *North Dublin: city and environs.* Dublin: M.H. Gill and Son.

Craig, Maurice J. (1977). *Classic Irish houses of the middle size.* London: Architectural Press.

D'Alton, John (1838). *The history of the county of Dublin.* Dublin: Hodges and Smith.

Dawson, T. (1976). The road to Howth, *Dublin Historical Record*, 29(4), 122–32.
Dublin City Architect's Blog. http://www.dublincityarchitects.ie/the-former-newcomen-bank-dublin-city-council-rates-office/ 19 August 2016. Accessed June 2023.
Foster, Roy (1992). *The Oxford history of Ireland*. Oxford: Oxford University Press.
Frame, Robin (1992). Commissions of the Peace in Ireland, 1302–1461. *Analecta Hibernica*, 35, 1–43. http://www.jstor.org/stable/25511057
Garrett, Arthur. (1970). *From age to age: history of the parish of Drumcondra, North Strand, St Barnabas*. Dublin: Blackrock printers.
Garrett, Arthur (2006, revised edition). *Down through the ages: the history of Killester church and parish 1926–2006*. Dublin, pamphlet produced for Killester parish.
Gillespie, Raymond (1996). *A history of Christ Church cathedral*, Dublin: Four Courts Press.
Gogarty, Claire (2013). *From village to suburb: the building of Clontarf since 1760*. Dublin: Clontarf Books.
Gwynn, Aubrey and Hadcock, R.N. (1970). *Medieval religious houses: Ireland*. London: Longman.
Gwynn, Lucius (1911). The life of St Lasair, *Ériu*, 5, 73–109.
Hamlin, Ann (1980). Two cross heads from County Fermanagh: Killesher and Galloon, *Ulster Journal of Archaeology*, 43, 53–8. http://www.jstor.org/stable/20567847
HMSO (1919). *Housing of the working classes in Ireland*. Dublin: HMSO.
Howard, Ebenezer (1898). *Tomorrow – a peaceful path to real reform*. UK: Swan Sonnenschein & Co. Reprinted as *Garden cities of tomorrow*, 1902.
Howlett, Liam (1979). The Killester charter, *Dublin Historical Record*, 32, 69–71.
ISSLT (1935). The Irish Sailors' and Soldiers' Land Trust. Accounts for the year ended 31 March 1935. London: HMSO.
Incorporated Council of Law Reporting for Ireland (1937). Bridget Casey and others v. The Irish Sailors' and Soldiers' Land Trust, defendants, *The Irish Reports*, 208, 208–24.
Irish Law Times Reports (1946). *Leggett & Others v. Irish Sailors' and Soldiers' Land Trust and The Attorney-General*, 80, 33ff.
Kelleher, Humphrey (2023). *A place to play, the people and stories behind 101 GAA grounds*. Dublin: Merrion Press.
Kenny, J. (1934). *History of Coolock parish*. Dublin: Olympia Printworks.
Landed Estates Court. Record of Proceedings. Vol. 1, Nov. 1858–1863. National Archives, Ireland.
Lavery Committee (1929). *Government memorandum on the report of the committee on claims of British ex-servicemen*. Dublin: Stationery Office.
Lefroy, Henry (1927). *History of Sailors' and Soldiers' Irish Land Trust (Ireland) 1915–1925*. Limerick: McKern and Sons Ltd.
Lennon, Colm (2018). *Clontarf. Irish historic towns atlas: Dublin suburbs, No. 1*. Dublin: Royal Irish Academy.
Lewis, Samuel (1837). *A topographical dictionary of Ireland*. London: S. Lewis and Co.
Loeber, R. (1981). *A biographical dictionary of architects in Ireland 1600–1720*. Ireland: J. Murray.
MacArthur, B. (ed.) (1992). *The Penguin book of twentieth-century speeches*. UK: Viking.

McCormack, Anthony M. (2004). Robert St Lawrence, second baron Howth. *Dictionary of Irish biography*. UK: Cambridge University Press.

McGrath, S. (2012). The lonely Thomas Gleadowe-Newcomen. https:// comeheretome.com/2012/11/19/the-lonely-thomas-gleadowe-newcomen/accessed 12 August 2023.

McManus, R. (1996). Public utility societies, Dublin Corporation and the development of Dublin, 1920–1940, *Irish Geography*, 29(1), 27–37.

McManus, R. (1999). The 'Building Parson' – the role of Reverend David Hall in the solution of Ireland's early twentieth-century housing problems. *Irish Geography*, 32(2), 87–98.

McManus, R. (2004). The role of public utility societies in Ireland, 1919–40. In *Surveying Ireland's past, multidisciplinary essays in honour of Anngret Simms*, edited by Howard Clarke, Jacinta Prunty and Mark Hennessy, 613–38. Dublin: Geography Publications.

McManus, R. (2002, 2021). *Dublin 1910–1940*. Volume 2 of The Making of Dublin City. Joseph Brady and Anngret Simms editors. Dublin: Four Courts Press.

Mills, James (1891). *Account roll of the priory of the Holy Trinity, Dublin, 1337–1346*. Dublin: Four Courts Press.

Murphy, Margaret and Potterton, Michael (2010). *The Dublin region in the Middle Ages: settlement, land-use and economy*. Dublin: Discovery Programme.

Murray, James (2009). *Enforcing the English reformation in Ireland. Clerical resistance and political conflict in the diocese of Dublin, 1534–1590*. UK: Cambridge University Press.

Natura Environmental Consultants (2005). Killester Graveyard Management Plan.

Ó Dufaigh, S. (2004). Lasair of Aghavea. *Clogher Record*, 18(2), 299–318.

Ó Maitiú, Séamas (2003). *Dublin's suburban towns, 1834–1930*. Dublin: Four Courts Press.

O'Neill, Michael (2011). *Bank architecture in Dublin: a history to c.1940*. Dublin: Dublin City Council.

Purves, G. (1987). The life and work of Sir Frank Mears: planning with a cultural perspective. PhD thesis, Heriot Watt University, Edinburgh.

Quinlivan, A. (2021). *Vindicating Dublin*. Dublin: Dublin City Council and Four Courts Press.

Rickard, Kevin (2017). Howth and its maritime past. *Dublin Historical Record*, 70(1), 19–35.

Rowley, Ellen (ed.) (2018). *More than concrete blocks, Volume 2, 1940–1972*. Dublin: Dublin City Council and Four Courts Press.

Sheehy, Maurice P. (1963/64) Unpublished medieval notitiae and epistolae. *Collectanea Hibernica*, 6/7, 7–17.

Simington, R.C. (1945). *The Civil Survey of 1654–1656*. Vol. VII. County of Dublin. Dublin: Stationery Office.

Stout, G. (2012). Review of *The Dublin region in the Middle Ages: settlement, land-use and economy*, by M. Murphy and M. Potterton. *Irish Economic and Social History*, 39, 153–4. http://www.jstor.org/stable/24338832

Taylor, George and Skinner, Andrew (1778). *Maps of the roads of Ireland, surveyed in 1777*. London and Dublin. Available online at www.askaboutireland.ie

Thom's Directory (various). *Thom's Irish Almanac and Official Directory of Ireland*. 1868, 1880, 1894, 1910, 1918, 1927, 1931. Dublin: Alex. Thom and Co.

Tudor Walters, J. (1918). *Report of the committee appointed by the president of the Local Government Board and the secretary for Scotland to consider questions of building construction in connection with the provision of dwellings for the working classes in England, Wales and Scotland.* London: HMSO.

Walsh, Robert (1888). *Fingal and its churches.* Dublin: William McGee.

Wright, G.N. (1825). *An historical guide to the city of Dublin.* London: Baldwin, Cradock and Joy.

Illustrations

1. Feoffment from Jenico (Preston) Viscount Gormanston and others, to Nicholas (St Lawrence) Lord Howth, of the lands of Killester, Baldongan, and other townlands in Co. Dublin, 1626 May 1. [D.9994 Howth Castle Papers. NLI] — 3
2. Killester church (in ruins), 1769. Gabriel Beranger. [PD 1958 TX_001(B). NLI] — 5
3. Extract from map of the County Dublin. William Petty, *Hiberniae delineatio*, 1683, showing 'Killaster' placename. [GE CC 1260. BNF] — 7
4. H.G. Leask watercolour of the Killester mansion house, c.1907. [AD2477. NLI] — 9
5. The environs of Killester. Extract from John Rocque, *A survey of the city, harbour, bay and environs of Dublin … 1757*. [PC] — 11
6. The road from Dublin to Howth. Map 148, depicting residence of 'Gleadowe Esq.'. George Taylor and Andrew Skinner, *Taylor and Skinner's Maps of the roads of Ireland, 1777*. [TCD] — 12
7. [A survey of] Part of Killester, Co. Dublin belonging to Richd. Cook Esq. 8th July 1775 [by] Thomas Sherrard, showing formal gardens to the rear of his residence and naming neighbouring landowners including W.G. Newcomen Esq., Mr Graham, Mr Dillon and Mr Heatly. [Longfield Collection MS 21 F. 51/(118). NLI] — 13
8. Portraits of the Gleadowe Newcomen family. Benjamin Wilson (1721–88), *Portrait of Sir William Gleadowe Newcomen seated at a table and the Hon. Thomas Newcomen as a child*, oil on canvas, 73.5cm x 61cm [TA]. Thomas Hickey, *Charlotte Newcomen, Lady Gleadowe Newcomen, 1st Viscountess Newcomen (c.1747–1817) with her daughters Jane, Teresa and Charlotte in a garden*, oil on canvas, c.1767–80. [NT] — 15
9. View of the entrance to Killester Demesne by Edward McFarland, 1853. [PD 1986 TX 8. NLI] — 18
10. The Newcomen Bank at Castle Street. [DCC] — 20
11. The ceiling in Newcomen Bank. [DCC] — 21
12. Gleadowe Newcomen tomb at Drumcondra churchyard. Photograph: Ruth McManus. — 21
13. Killester House. Extract from Ordnance Survey plan, 1:10,560, sheet 19, 1837 edition. — 25
14. Portraits of Thomas Popham Luscombe and Catherine Tooke Robinson Luscombe upon their marriage, by Benjamin Delacour. [FA] — 27
15. Primary (Griffith) valuation for Killester Demesne – the summary and the surveyor's maps. [Ask About Ireland] — 29
16. Extracts from Landed Estates Court records for Killester, 1863. [Find My Past] — 31–2
17. Parish map, 1863. Extract from Ordnance Survey plan, 1:2,500, Co. Dublin sheet XIX(1), 1863 edition. [FSLA] — 33
18. Index map of Dublin North properties. Howth estate. [FSLA] — 34
19. Killester North estate map. Howth estate. [FSLA] — 36–7

20	Killester South estate map. Howth estate. [FSLA]	38–9
21	Venetian window joinery salvaged from Killester House and used to frame a fireplace in the entrance hall at Howth Castle, *c.*1909. Photograph: Rob Goodbody.	43
22	Chimneypiece salvaged from Killester House and reused in Lutyen's new library at Howth Castle, *c.*1909. [DCC]	43
23	First World War recruitment poster. John Redmond exhorting men to join up. [PC]	47
24	One suggested layout for working class houses in Ireland. [HMSO, 1919]	49
25	Present-day aerial view of housing in Hampstead Garden Suburb. [Google Earth]	51
26	Killester, Garden and Rural Suburb. Map of the Killester area showing the land owned by the Dublin Garden Estates Company, outlined in light blue. [OPW/5HC/4/973. NA]	53
27	Killester Demesne near Dublin. Suggested Garden Suburb. Birds-Eye view from South. [PRIV1232/1. NA]	58
28	Killester Garden Village Near Dublin. General plan showing proposed water scheme. [OPW/5HC/4/973. NA]	63
29	Plan for Type G2 bungalow. [OPW/5HC/4/973. NA]	64
30	The final shape of Killester Garden Village showing the later infill. Ordnance Survey plan, 1:2,500, sheets 15(13) and 19(1), 1938 edition. [PC]	66
31	St Brigid's church, Killester, *c.*1943. [ITA Coolock-03. DCLA]	71
32	The DUTC and Contemptible Omnibus Company racing towards a passenger. *Dublin Opinion*, November 1925: 677. [PC]	73
33	A DUTC bus and the developing shopping area in Killester, 1920s. [PC]	74
34	Advertisement for the Contemptible Bus Company, showing its diversification into regional routes. Dublin's Transport Network, *3d. Motor News*, October 1928. [PC]	74
35	Killester and environs, early 1930s. Extract from Ordnance Survey map, 1:20,000, Dublin and Environs, Provisional Popular Edition, Sheet 265b, 1933 edition. [PC]	75
36	Killester and environs, late 1940s. Extract from Ordnance Survey map, 1:25,000, Dublin Popular Edition, 1948 edition. [PC]	76
37	Killester and environs, late 1950s. Extract from Ordnance Survey map, 1:25,000, Dublin Popular Edition, 1959 edition. [PC]	77
38	Layout of Seafort Gardens. Ordnance Survey plan, 1:2,500, sheets 18(12) and 18(16), 1939 edition. [PC]	78
39	Seafort Gardens at completion. [Joseph Brady and G. & T. Crampton photograph archive, UCD]	79
40	Housing on Fairhill Road Upper, Claddagh, Galway. [AP7/171. TNA]	79
41	Layout of Rosary Gardens, Library Road. Ordnance Survey plan, 1:2,500, sheet 23(6), 1939 edition. [PC]	80
42	Layout of the ISSLT housing scheme in the Drumcondra reserved area. Ordnance Survey plan, 1:2,500, sheets 14(15) and 18(3), 1938 edition. [PC]	82
43	ISSLT crest on the façade of a house on Lambay Road, Drumcondra. [PC]	83

Illustrations 133

44 Layout of the Milltown scheme between Churchtown Road and the Harcourt Street railway line. Ordnance Survey plan, 1:2,500, sheets 22(3) and 22(7), 1938 edition. [PC] 84
45 Layout of the Kimmage (Larkfield) scheme. Ordnance Survey plan, 1:2,500, sheet 22(2), 1938 edition. [PC] 85
46 ISSLT at Quarry Road, Beggsboro, on the edge of the Dublin Corporation Cabra housing area. Ordnance Survey plan, 1:2,500, sheets 18(2) and 18(6), 1943 edition. [PC] 86
47 The entrance piers to one of the culs-de-sac on Quarry Road. [PC] 87
48 The Ballinteer Gardens scheme comprising terraced two-storey houses and bungalows. Ordnance Survey plan, 1:2,500, sheet 22(12), 1936 edition. [PC] 88
49 The cul-de-sac at Ballinteer Gardens. [AP7/171. TNA] 89
50 One of the bungalows at the entrance to Ballinteer Gardens. [AP7/171. TNA] 89
51 Park Villas at Castleknock. Ordnance Survey plan, 1:2,500, sheet 18(1), 1948 edition. [PC] 90
52 Killester and its wider environs. Extract from Ordnance Survey Popular Edition, 1:25,000. [PC] 104
53 Killester and District with the census areas shown. Extract from Ordnance Survey Popular Edition, 1:25,000, 1948 edition. [PC] 106
54 Aerial view of new Collins Avenue housing schemes and Killester in 1952. [XAW044955. HES] 108
55 Aerial view of new Collins Avenue housing schemes and Killester in 1952, north-west. [XAW044960. HES] 109
56 Aerial view of Killester Park under construction in 1952. [extract from image XAW044955. HES] 110
57 The former water tower at the ex-servicemen's estate. [extract from image XAW044960. HES] 110
58 Killester House seen between the railway line and Killester Avenue. [extract from image XAW044955. HES] 111
59 The sale of Killester House in 1937. *Irish Times,* 2 March 1937:16. 112
60 Extract showing the surviving Killester House [XAW044960. HES] 113
61 The nun's walk in 1943. [ITA-Coolock-04. DCLA] 113
62 The 'Killer' shortly before its demolition. [DCC] 116
63 The Ordnance Survey Popular Edition, published in 1959. [PC] 117
64 Killester and environs in the early 1970s. [PC] 119
65 Holy Faith convent in 1943. [ITA-Coolock-05. DCLA] 123
66 Killester and environs in a recent image captured by Google Earth. 124

We express our thanks to the individuals and organisations who have licenced the use of images used in the text. A guide to the abbreviations used follows. All copyrights are acknowledged.

BNF	Licenced from Bibliothèque Nationale de France
DCC	Courtesy of Dublin City Council
DCLA	Courtesy of Dublin City Library and Archives
FA	Courtesy of Freeman's Auctioneers
FLSA	Courtesy of Fingal Local Studies and Archives
HES	Historic Environment Scotland
NA	Reproduced with kind permission of the Director of the National Archives
NLI	Reproduced courtesy of the National Library of Ireland
NT	© National Trust UK
TA	Courtesy of Tennants Auctioneers, North Yorkshire
TCD	Courtesy of the Board of Trinity College Dublin
TNA	Licenced from The National Archives (UK)

Other images are from private collections (PC).

Index

abbey, 23, 26
Abbeyfield, 23, 26, 62, 66, 75–6, 119
Abercrombie, Patrick, 52, 54
Aberdeen, Lord, 51
Acts, 48–9, 60, 67–8, 70, 72, 77, 97–8
advertisements, 13–14, 26, 28, 56–7, 64, 74, 107, 111
aerial photography, 51, 107–11
agriculture, 4, 42, 45, 48–9, 57, 87, 99, 101, 106
allotments, 48, 57, 59
Amiens Street station, 70
Anglo-Normans, 4, 6
annual rent, 4, 12, 31, 59
apartments, 121, 123, 125
Árd Lorcain Garden Village, 56
Armistice Day, 70
arrears, rent, 91, 93
Artane, 24, 35, 44, 56, 75, 100
Artane House Building Estate, 100
Aston, E.A., 51–2, 54–6, 59–60, 67

Baldoyle, 35
Ballinteer Gardens, 88–9
banks, 11, 16–17, 19–20, 22
    private, 14, 16
bankers, 10, 16
banking, failures, 16, 20
barns, 4, 30, 107
bathroom, 19, 62, 69, 81
bawn, 8–9
bay and environs of Dublin, 10–11, 13, 23
Belfast, 68–70, 91
Beranger, Gabriel, 5–6
blocks, 81, 121, 123
Bolshevism, 92

boundaries, 4, 11, 41, 62, 66, 82, 100, 118–19
    borough, 59
    nineteenth-century demesne, 10
bricks, 30, 86, 125
    red, 85, 87
Brigid, Saint, 1–2
British army, 46, 57
British government, 67, 69, 93, 97–8
British Legion (Royal), 70–1, 91–2, 97
Brookwood Avenue, 117–19
Brookwood Court, 125
builders, 17, 42, 57, 64–5, 81, 83, 99, 102
building contractors, 57, 64
building cost, 68, 105
building land, 61, 102–3
Bull Wall, 108
bungalows, 62, 88–9, 107
buses, 72–3, 107
business, 16, 19, 24, 28, 57, 60, 72, 98, 114

Cabra, 60, 87, 89, 96
Canon David Hall, 55
Carrigglas, 14, 19
Carrigglas Manor, 17
Castle Avenue, 55, 67, 119
Castleknock, 88, 90
Castle Street, 14, 16, 20, 22, 24
cattle, 45, 100, 102
ceilings, 16–17, 21
census, 8, 18, 28, 41–2, 99–100, 106, 117–21
census areas, 106, 117, 119
Chancery Court, 22
chapel, 2, 5–6, 23, 115
charters, 2–4

135

children, 22, 48, 70, 95, 101, 114, 116
Christ Church, 3–8
church, 1–6, 23, 26, 35, 67, 72, 101
   medieval, 5, 10
   ruined, 5–6, 23, 26
Church of Ireland, 23, 42
Churchtown, 84–6
cinema, 116–17, 120
city architect, 52, 54–5
city boundaries, 44, 55, 103
city centre, 72, 102, 115
City Hall, 16, 102
Civic Exhibition, 51
Civics Institute, 54
Civil Survey, 5–6, 8–9, 12
Civil War, 45, 52
Claddagh, 79, 81
Clanawley Road, 105, 111
class, 42, 69, 92, 102
Clontarf, 10, 14, 19, 24, 31, 35, 59, 106–7, 118, 120, 122
   Golf Club, 102–3
Clontarf, Town Hall, 70
Clontarf East, 100, 106–7, 117–19, 122
Clontarf West, 41–42, 100, 106, 117–20, 122
Collins Avenue, 2, 55, 73, 75, 103–6, 115, 117, 120, 123
colonies, 49–50, 68, 71, 93
commissioners, 82–83
committee, 47, 52, 54–5, 57, 60–1, 65, 92–3, 96–7
committee of inquiry, 96
competition, 50, 52, 72–3
Congested Districts Board, 48
Coote, Chidley, 9–10
construction, 16, 26, 62, 64–5, 99, 102, 107, 110, 121, 123, 125
Contemptible (Omni)Bus Company, 72–4
contractor, 65, 72, 80–1, 86, 115
convent, 1–2, 26, 115, 123
   grounds, 117
   in ruins, 2, 25

Coolock, 1–2, 4, 8, 28, 35
co-partnership societies, 55
cottages, 31, 35, 42, 60–6, 69, 71, 78, 94–5, 97–8, 102, 105
creditors, 8, 22
Cromwellian era, 2, 8
Crumlin, 102, 115
culs-de-sac, 87–8, 122

debts, 12, 19, 22, 55
demesne, 11–12, 19, 23, 25–26, 30, 42, 44–5, 55, 62, 75–6, 101, 119
   house, 35, 107, 110
   lands, 10, 22, 26, 32, 40
demolition, 44, 116
design, of houses, 16–17, 19, 24, 50, 54–5, 59, 66, 72, 81, 85–7
discrimination, 92
Dollymount, 35, 106–7
Donnycarney, 4, 60, 72–3, 75, 90, 101–3, 105, 107, 115, 118
   House, 103
   scheme, 102, 105, 118–20
Down Survey, 8
Drogheda Railway Company, 26, 30–1, 35
Drumcondra, 4, 17, 22, 41, 60, 72, 82–3, 87, 103
   churchyard, 21
   Rural, 41–2
Dublin Castle, 16, 103
Dublin City Council, 17, 123
Dublin Corporation, 44–5, 50, 52, 55, 57, 59–62, 65, 75, 82–3, 98–9, 101–3, 105, 107, 114, 117–18, 122
   housing committee, 59–61
   housing department, 55, 59
   report, 52, 57, 59, 83, 102, 105
Dublin Evening Mail, 22
Dublin Evening Post, 22, 24
Dublin Evening Telegraph, 42, 55
Dublin Garden Estates Company, 53–4, 56
Dublin Opinion, 73
Dublin Popular Edition, 73, 76–7

Index 137

Dublin United Tramways Company, *see* DUTC
Dunluce Road, 100, 119
DUTC (Dublin United Tramways Company), 72–3
dwelling-house, 12–13, 18–19, 30
dwellings, 26–7, 30, 35, 41–2, 47, 81, 85, 88, 102, 111, 118

earthwork, 4, 11
Easra, 1–2, 125
economy, 6, 56, 65, 69, 72
employment, 10, 46, 92
entrance, 16, 18, 40, 88–9
    formal, 19, 87, 89
entrance hall, 26, 44
entrance pillars, 87–8
estates, 4, 10, 19, 22, 26, 31, 75, 77, 93, 95, 101
    troublesome, 77
Evening Freeman, 40
Evening Herald, 28, 56, 101, 103, 107, 116–17, 121
Evening Irish Times, 42, 44
eviction, 93

facades, 83, 85–7, 89
Fade Street, 10
Fairbrothers' Fields, 52, 60, 65, 90
Fairview Co-Operative Tillage Society, 60
families, 20, 22, 26, 41–2, 102–3, 107, 118
    large, 13, 40, 42, 118
farmer, 24, 41–2, 48, 102
farms, 30, 45, 48, 60, 103, 105, 123
field boundaries, 10–11
fields, 8, 11, 17–19, 25, 40, 62, 93
First World War, 46–7, 52, 57
fixed baths, 62, 118
freehold, 4, 111
Freeman's Journal, 28, 40–1, 45, 61, 103
Free State, 67–8, 70, 85, 91–2, 94–5, 99, 103

fruit, 28, 46, 103
Furry Park House, 100, 104
Furry Park Road, 119

Gandon, James, 17–19
garden city / suburbs, 45, 47, 50, 52, 54–6, 62, 65, 75
gardens, 12–15, 17–19, 23–7, 32, 35, 56, 62, 67, 100, 105, 123
    large, 13, 44, 87, 107
garden suburb, first, 50
garden village, 45, 56, 62
gate lodges, 18–19, 26, 35
gates, 17, 19
Geddes, Patrick, 50, 52
Gleadowe Newcomen family, 12, 14–16, 19, 21, 26
grape-house, 17, 19, 26
graveyards, 5–6, 26, 35, 67
Greater Dublin Reconstruction Movement, 52, 54
greenhouses, 17, 22, 26, 100, 111
Griffith, Richard, 28–9, 54, 59
Griffith's Valuation, 18, 28, 30

Harcourt Street railway line, 84–5
Harmonstown, 42, 117
Hibernian Bank, 20, 120
High Court, 94–5
HM Treasury, 65, 68, 77, 91
Holy Faith order, 101, 123
home rule, 46, 48
horses, 13, 24, 114
housing schemes, 44, 69, 94, 114
    distinctive, 45
housing tenure, 118, 120
Howth, 2, 6–10, 12, 17–18, 24, 30–1, 41, 44
    Castle, 3, 19, 43–4
    estate, 34–6, 38, 42, 44, 57
Howth, lords of, 2, 6, 8, 10, 41
Howth Road, 2, 18–19, 24, 26, 67, 72, 75, 99–101, 105–6, 114, 119–21

ice house, 25, 101
Independent Newspapers, 57
infill, 2, 66, 121–3
Irish Builder and Engineer, 55, 61, 65, 67
Irish Free State, 67
Irish government, 97–8
Irish Independent, 41, 61, 73, 99, 102, 107, 115–17, 121–2
Irish Press, 56, 101, 105, 107, 116
Irish Times, 28, 45–6, 51–2, 54–6, 64–5, 70–3, 78, 80–1, 91–3, 99–100, 102–3, 107, 111–12, 114–15, 121, 125

Johnston, Justice 94

Kilbride Road, 99, 102, 104, 119
Killester Abbey, 24–6, 30, 35, 44, 107, 110
Killester and environs, 41, 53, 75–7, 100, 102, 106, 118–19, 121, 125
Killester Avenue, 2, 5, 10, 19, 100, 105–7, 111–12, 114–15, 119, 121
Killester Cinema, 115–16, 123
Killester Demesne, 10, 17–19, 25–6, 29–30, 35, 40–2, 44, 57, 59, 62, 107, 125
Killester depot, 114
Killester Gardens, 26–8, 32, 35, 102, 107, 111–13
Killester Garden Village, 46–7, 49, 51, 55, 57, 59, 61–2, 65–97
Killester Hall, 12, 40–1
Killester House, 1, 11, 22–3, 25–7, 30, 35, 42–5, 101–2, 107, 112, 117, 121
Killester Lane, 28, 42, 102, 105
Killester Lodge, 14, 24, 35
Killester manor house, 10, 26
Killester mansion house, 9, 14, 17, 44, 59, 67
Killester North, 25, 29–30, 35–6, 40–2
Killester Park, 26, 30, 99–100, 103–5, 107, 110, 117, 119
Killester South, 26, 28, 30, 35, 38, 40–1
Kimmage, 85, 87
Kingstown, 24, 80–1

Landed Estates Court, 12, 17, 30–1
landlords, 9, 105
Larkfield, 85, 87
Lavery Committee, 70, 91–2
law agents, 60–1
Lefroy, Henry, 46, 67–8, 70, 92
Leggett, Robert, 94
Legion Hall, 45, 67, 71, 110
legislation, 47–8, 50, 57, 67–9, 72, 92, 94, 97–8
Library Road, 80–1
Lipton, 120
local authorities, 57, 68, 98, 120, 122
Local Government, 65, 68, 73, 96–7, 121
Local Government Board (LGB), 49–50, 52, 55–6, 59–62, 64–5, 67, 78
Local Representative Relief Committee, 57
London, 24, 26, 50
lord mayor of Dublin, 52, 61, 71, 112, 115
Luscombe (family), 23–4, 26, 30

Mahon, Sir Bryan, 68, 92
maintenance, 91, 93, 95–6, 98
Malahide Road, 35, 42, 55, 99–100, 102, 105–6, 109
maps, 7, 10–12, 25–6, 30, 32, 35–6, 38, 40, 53, 66–7, 119
Marino, 52, 60, 64–5, 73, 82, 90, 100, 106–7, 115
  House, 44
market gardening, 27, 100, 113
McLaughlin, Sir Henry, 52, 54, 57, 61, 92
meadows, 3, 8, 13, 18–19
Mears, Frank, 51–2, 54, 59, 62, 64
Merrion Square, 92
Middle Third, 10, 45, 62, 71, 76, 96, 119
Milltown, 81, 84–85, 87
minister for local government, 68, 96–7, 121
money, 14, 16, 29, 61, 68–70, 81, 97–8
Morning Post, 20
motte, 4, 8
Murnaghan, Justice 94–5

Index

National Archives, 28, 54, 59, 62, 68, 77
Newcomen (family), 12–17, 20, 22, 26
Newcomen
   Bank, 16, 20–1
   Bridge, 17
   Terrace, 26
newspapers, 56, 70, 81
North Dublin Rural District Council, 59
Northern Ireland, 68, 91, 95, 97–8
North Strand, 17, 115
nuns, 23, 110

Oireachtas, 68, 92, 98
Old Killester Tenants Rights, 96
orchards, 66, 89, 111–12, 125
Orchard Road, 75–76
Ordnance Survey, 1, 24, 73, 75–77
   first edition, 18, 26
Ordnance Survey plan, 25, 33, 66, 78, 80–90
Ordnance Survey Popular Edition, 104, 106, 117, 119
overcrowding, 102, 106, 118

parish, 2, 8, 17, 23, 30, 40, 115
   map, 33, 35
Park, Victoria, 102–3
parkland, 19, 25–6, 35
Park Villas, 88, 90
parlour, 19, 62, 81, 85
Peck's Lane, 88
piggery, 41–2
planning, 50, 54, 59, 101
   cities and town, 50, 60
   civic, 51
plans, 16–17, 44, 50, 52, 54, 59–60, 62, 64
plotholders, 59–60
plots, 12, 35, 49, 55–7, 59, 67, 100
population, 8, 40, 67, 71, 99–100, 102, 106, 114, 118, 122
   growing, 100, 105–6, 114–15
Premier dairies, 114–15
purchase schemes, 96–7, 107

Quarry Cottages, 42, 102
Quarry Road, 87

Raheny, 4, 10, 24, 26, 28, 35, 40, 71, 88, 100, 107
railway line, 26, 30, 40, 62, 72, 87, 102, 109, 111, 121
recruitment for British army, 46, 48, 57
rent, reductions, 93–94
rent levels, 70
rents, 14, 32, 61, 65, 70, 77, 90–1, 93–6, 105
   Killester, 90–1
   levy, 97
   protest, 91
roads, 3, 12, 18, 50, 75, 81, 87, 102–3, 105
   main, 55, 67, 73, 86
Rocque, John, 10–11, 25
Roman Catholic, 2, 42, 72, 92
roof, 9, 81, 87, 107
   slated, 99
rooms, 16, 30, 41–2, 56, 67, 81, 85
Rosary Gardens, 80
Royal Canal, 17
Royal Canal Company, 17

Sackville Street, 24, 28, 57
Sallynoggin, 81, 88
Saorstát Éireann, 93
Saunders' Newsletter, 12–14, 19, 26, 28
schemes (housing), 52, 55, 57, 59, 62, 64–6, 69, 78, 80–1, 85–7, 97–8, 101–2, 105, 125
school, secondary, 115, 123
Seafort Gardens, 78–80
Seanad Éireann, 65
Second World War, 97, 99–101, 105, 115
semi-detached
   bungalow, 78
   houses, 56, 59, 62, 87, 100
servants, 10, 42, 100
services, 40, 65, 72–3, 93, 114
sewers, 83, 105

shopping parade, 105, 114, 120
shops, 18, 42, 55, 100, 114, 116–17, 120
    residential, 107, 114
short terraces, 83, 86–7, 89, 105
slums, 47–8, 61
soldiers, 48–9, 61, 94
South Field, 62
spaces, open, 50, 54, 66–7, 105, 107, 125
St Brigid's boys' national school, 101
St Brigid's church, 71
St Brigid's girls' national school, 125
St Brigid's Road, 115, 119, 122
St Lawrence Road, 72, 106
stones, 4, 30, 101
    foundation, 72, 107
suburbs, 1, 47, 52, 54–6, 59, 106, 114
supermarkets, 116, 120
Supreme Court, 94–5

tenancy, 27, 59, 95
tenant, purchasers, 71, 105, 118
tenants, 22, 32, 35, 40–1, 49, 76–7, 90–1, 93–8, 100, 118, 125
    Killester, 90, 96
tenders, 82, 89, 115
tenements, 102, 105
tennis court, 111
tenure, 48, 120, 122
Terenure, 87, 102
Thom's Directory, 40–1, 75, 114, 121
tithes, 23–4

toilet, 22, 105, 118
town planning, 50–2, 57
trains, 70, 72
trees, 17, 19, 25–6, 50, 54, 101, 107, 110
trustees, 31, 69, 77, 98

unemployment, 57, 64, 122
United Kingdom, 14, 47–8, 57, 67–8, 70, 77, 91, 98
Unwin, Raymond, 50, 52
urban area, 68–9
utility societies, public, 55–6, 60, 82–3, 107

vegetables, 13, 28, 103, 125
Venetian Hall, 26, 101, 117, 121–2
Vernon Avenue, 55–6, 67, 119

walls, 5, 8, 10, 45, 111, 113
    bawne, 12
    defensive, 8, 12
    high brick, 6, 13, 111, 113, 116
war, 46–8, 57, 59, 94
    memorial, 70, 92
wards, 100, 106, 118, 120
water, 62, 64, 81, 83, 98
    tower, 110
widows, 28, 42, 76, 93, 95–8
wooded area, 25, 35
woods, 3, 8, 101
Woodville, 24, 41
working classes, 47, 50, 57, 90